George Abbott Osborne

An Elementary Treatise on the Differential and Integral Calculus

George Abbott Osborne

An Elementary Treatise on the Differential and Integral Calculus

ISBN/EAN: 9783337811518

Printed in Europe, USA, Canada, Australia, Japan

Cover: Foto ©berggeist007 / pixelio.de

More available books at **www.hansebooks.com**

AN

ELEMENTARY TREATISE

ON THE

DIFFERENTIAL AND INTEGRAL CALCULUS,

WITH EXAMPLES AND APPLICATIONS.

BY

GEORGE A. OSBORNE, S.B.,

PROFESSOR OF MATHEMATICS IN THE MASSACHUSETTS
INSTITUTE OF TECHNOLOGY.

D. C. HEATH & CO., PUBLISHERS
BOSTON   NEW YORK   CHICAGO
1903

COPYRIGHT, 1891,
BY GEORGE A. OSBORNE.

PREFACE.

THIS work, intended as a text-book for colleges and scientific schools, is based on the method of limits, as the most rigorous and most intelligible form of presenting the first principles of the subject. The method of limits has also the important advantage of being a familiar method; for it is now so generally introduced in the study of the more elementary branches of mathematics, that the student may be assumed to be fully conversant with it on beginning the Differential Calculus.

The rules or formulæ for differentiation in Chapter III. differ in one respect from those in similar text-books, in being expressed in terms of u instead of x, u being any function of x. They are thus directly applicable to all expressions, without the aid of the usual theorem concerning a function of a function.

After acquiring the processes of differentiation, the student in Chapter V. is introduced to the differential notation, as a convenient abbreviation of the corresponding expressions by differential coefficients. This notation has manifest advantages in the study of the Integral Calculus and in its applications.

In Chapter IX. and subsequent pages I have introduced for Partial Differentiation the notation $\frac{\partial}{\partial x}$, which has recently come into such general use.

The chapters on Maxima and Minima have been placed after the applications to curves, as the consideration of that subject is much simplified by representing the function by the ordinate of a curve. Maxima and Minima may be taken, if desired, with equal advantage immediately after Chapter XIII.

In Chapter X., Integral Calculus, I have taken the problem of finding the Moment of Inertia of a plane area, as a better illustration of double integration than that of finding the area itself. The student more readily comprehends the independent variation of x and y in the double integral,

$$\iint (x^2 + y^2)\, dx\, dy, \quad \text{than in} \quad \iint dx\, dy.$$

A few pages of Chapter XII., Integral Calculus, are devoted to a description of the Hyperbolic Functions together with their differentials, and a comparison is made with the corresponding Circular Functions.

G. A. OSBORNE.

Boston, 1895.

CONTENTS.

DIFFERENTIAL CALCULUS.

CHAPTER I.
FUNCTIONS.

ARTS.		PAGES.
1–4.	Definition and Classification of Functions	1, 2
5.	Notation of Functions. Examples	3, 4

CHAPTER II.
DIFFERENTIAL COEFFICIENT.

6, 7.	Limit. Increment	5
8–10.	Differential Coefficient. Examples	6–9

CHAPTER III.
DIFFERENTIATION.

11–13.	Differentiation of Algebraic Functions. Examples	10–21
14–16.	Differentiation of Logarithmic and Exponential Functions. Examples	21–27
17, 18.	Differentiation of Trigonometric Functions. Examples	27–32
19, 20.	Differentiation of Inverse Trigonometric Functions. Examples	32–37
21, 22.	Differentiation of Inverse Function and Function of a Function. Examples	37–40

CHAPTER IV.
SUCCESSIVE DIFFERENTIATION.

23, 24.	Definition and Notation	41
25.	The nth Differential Coefficient. Examples	42–45
26.	Leibnitz's Theorem. Examples	45–47

Chapter V.

DIFFERENTIALS.

Arts.		Pages.
27.	Differentials as related to Differential Coefficients	48, 49
28.	Differentiation by Differentials	49
29.	Successive Differentials. Examples	50, 51

Chapter VI.

IMPLICIT FUNCTIONS.

30. Differentiation of Implicit Functions. Examples........ 52–54

Chapter VII.

EXPANSION OF FUNCTIONS.

32–36.	Maclaurin's Theorem. Examples	55–60
37–41.	Taylor's Theorem. Examples	60–63
42–45.	Rigorous Proof of Taylor's Theorem	64, 65
46–49.	Remainder in Taylor's and Maclaurin's Theorems	66–68

Chapter VIII.

INDETERMINATE FORMS.

50, 51.	Limiting Value of a Fraction	69
52, 53.	Evaluation of $\frac{0}{0}$. Examples	70–73
54–57.	Evaluation of $\frac{\infty}{\infty}$, 0∞, $\infty - \infty$. Examples	73–76
58.	Evaluation of Exponential Forms. Examples	76–78

Chapter IX.

PARTIAL DIFFERENTIATION.

59, 60.	Partial Differential Coefficients of First Order. Examples	79, 80
61–63.	Partial Differential Coefficients of Higher Orders. Examples	80–82
64, 65.	Total Differential of Functions of Several Variables. Examples	82–84
66.	Condition for an Exact Differential. Examples	85
67.	Differentiation of Implicit Functions	86
68, 69.	Taylor's Theorem for Several Variables	87, 88

CHAPTER X.

CHANGE OF VARIABLES IN DIFFERENTIAL COEFFICIENTS.

Arts.		Pages.
70.	Changing from x to y	89
71, 72.	Changing from y to z	90
73.	Changing from x to z. Examples	90–92

CHAPTER XI.

REPRESENTATION OF VARIOUS CURVES.

| 74–85. | Rectangular Co-ordinates .... | 93–98 |
| 86–93. | Polar Co-ordinates .... | 98–102 |

CHAPTER XII.

DIRECTION OF CURVE. TANGENT AND NORMAL. ASYMPTOTES.

94–97.	Direction of Curve. Subtangent and Subnormal. Examples	103–108
98, 98½.	Differential Coefficient of the Arc................	108, 109
99.	Equation of the Tangent and Normal. Examples ...	109–112
100–106.	Asymptotes. Examples	112–116

CHAPTER XIII.

DIRECTION OF CURVATURE. POINTS OF INFLEXION.

| 107–109. | Direction of Curvature........................... | 117 |
| 110. | Points of Inflexion. Examples................... | 118, 119 |

CHAPTER XIV.

CURVATURE. CIRCLE OF CURVATURE. EVOLUTE AND INVOLUTE.

111–113.	Definition of Curvature; Uniform and Variable	120, 121
114, 115.	Radius of Curvature. Examples	121–124
116.	Centre of Curvature	124, 125
117–121.	Evolute and Involute. Examples.................	125–128

CHAPTER XV.

ORDER OF CONTACT. OSCULATING CIRCLE.

Arts.		Pages.
122, 123.	Consecutive Common Points	129, 130
124, 125.	Osculating Curves	130, 131
126–128.	Analytical Conditions for Contact	131–133
129, 130.	Osculating Circle. Examples	133–136

CHAPTER XVI.

ENVELOPES.

131–133.	Series of Curves. Definition of Envelope	137, 138
134–136.	Equation of Envelope	138–140
137.	Evolute, the Envelope of Normals. Examples	140–144

CHAPTER XVII.

SINGULAR POINTS OF CURVES.

138–141.	Multiple Points	145–148
142, 143.	Points of Osculation. Cusps	149, 150
144.	Conjugate Points. Examples	150–152

CHAPTER XVIII.

MAXIMA AND MINIMA OF FUNCTIONS OF ONE INDEPENDENT VARIABLE.

145–149.	Definition. Conditions for Maxima and Minima derived from Curves	153–157
150, 151.	Conditions for Maxima and Minima by Taylor's Theorem. Examples	157–162
	Problems in Maxima and Minima	162–164

CHAPTER XIX.

MAXIMA AND MINIMA OF FUNCTIONS OF SEVERAL INDEPENDENT VARIABLES.

152–155.	Definition. Conditions for Maxima and Minima by Taylor's Theorem. Examples	165–171

INTEGRAL CALCULUS.

Chapter I.
ELEMENTARY FORMS OF INTEGRATION.

Arts.		Pages.
1, 2.	Definition of Integration. Elementary Principles	173, 174
3.	Fundamental Integrals	175, 176
4–7.	Derivation and Application of Fundamental Formulæ. Examples	176–187

Chapter II.
INTEGRATION OF RATIONAL FRACTIONS.

8, 9.	Preliminary Operation. Factors of Denominator	188, 189
10.	Case I. Examples	189–191
11.	Case II. Examples	191, 192
12.	Case III. Examples	192–195
13.	Case IV. Examples	195–198

Chapter III.
INTEGRATION BY RATIONALIZATION.

14–16.	Fractional Powers of x and of $a + bx$. Examples	199–201
17.	Fractional Powers of $a + bx^2$. Examples	201, 202
18, 19.	Expressions containing $\sqrt{\pm x^2 + ax + b}$. Examples	202–204
20.	Integration by Substitution. Examples	204, 205

Chapter IV.
INTEGRATION BY PARTS. INTEGRATION BY SUCCESSIVE REDUCTION.

21.	Integration by Parts. Examples	206–208
22–24.	Formulæ of Reduction. Examples	208–214

Chapter V.
TRIGONOMETRIC INTEGRALS.

25–27.	Integration of $\tan^n x\, dx$; of $\sec^n x\, dx$; of $\tan^m x \sec^n x\, dx$. Examples	215–218
28, 29.	Integration of $\sin^m x \cos^n x\, dx$. Examples	219–222

CONTENTS.

ARTS.   PAGES.

30. Trigonometric, transformed into Algebraic, Integrals. Examples 222–224
31, 32. Trigonometric Formulæ of Reduction. Examples..... 224–226
33–35. Integration of $\dfrac{dx}{a+b\sin x}$ and $\dfrac{dx}{a+b\cos x}$; of $e^{ax}\sin nx\,dx$ and $e^{ax}\cos nx\,dx$. Examples 226–229

CHAPTER VI.
INTEGRALS FOR REFERENCE.

36. Integrals containing $\sqrt{a^2-x^2}$; $\sqrt{x^2\pm a^2}$; $\pm ax^2+bx+c$. 230–235

CHAPTER VII.
INTEGRATION AS A SUMMATION. DEFINITE INTEGRALS.

37–40. Integration, the Summation of an Infinite Series 236–240
41–43. Definition of Definite Integral. Examples............ 240–244

CHAPTER VIII.
APPLICATION OF INTEGRATION TO PLANE CURVES. APPLICATION TO CERTAIN VOLUMES.

44–47. Areas of Curves. Examples 245–249
48, 49. Lengths of Curves. Examples 249–252
50, 51. Surfaces of Revolution. Examples 252–255
52. Other Volumes. Examples 255–257

CHAPTER IX.
SUCCESSIVE INTEGRATION.

53–56. Double and Triple Integrals. Examples 258–260

CHAPTER X.
DOUBLE INTEGRATION APPLIED TO PLANE AREAS AND MOMENT OF INERTIA.

57–60. Double Integration. Rectangular Co-ordinates. Examples 261–264
61–63. Double Integration. Polar Co-ordinates. Examples .. 264–266

Chapter XI.

SURFACE AND VOLUME OF ANY SOLID.

ARTS.		PAGES.
64, 65.	Area of any Surface. Examples	267–270
66, 67.	Volume of any Solid. Examples	270–273

Chapter XII.

HYPERBOLIC FUNCTIONS. CYCLOID, EPICYCLOID, AND HYPOCYCLOID. INTRINSIC EQUATION OF A CURVE.

69–71.	Definitions of Hyperbolic, and Inverse Hyperbolic, Functions	274–276
72, 73.	Differentiation of Hyperbolic Functions. Inverse Hyperbolic Functions as Integrals	276, 277
74, 75.	Hyperbolic Functions and the Hyperbola. Exercises	278–280
76–82.	Equation and Properties of the Cycloid	280–284
83–89.	Equations and Properties of the Epicycloid, and Hypocycloid	284–288
90–93.	Intrinsic Equation of a Curve and of its Evolute. Examples	289–292

DIFFERENTIAL CALCULUS.

CHAPTER I.

FUNCTIONS.

1. *Definition of a Function.* When the value of one variable quantity so depends upon that of another, that any change in the latter produces a corresponding change in the former, the former is said to be a *function* of the latter.

For example, the area of a square is a function of its side; the volume of a sphere is a function of its radius; the sine, cosine, and tangent are functions of the angle; the expressions

$$x^2, \quad \log(x^2+1), \quad \sqrt{x(x+1)},$$

are functions of x.

A quantity may be a function of two or more variables. For example, the area of a rectangle is a function of two adjacent sides; either side of a right triangle is a function of the two other sides; the volume of a rectangular parallelopiped is a function of its three dimensions.

The expressions

$$x^2 + xy + y^2, \quad \log(x^2+y^2), \quad a^{x+y},$$

are functions of x and y.

The expressions

$$xy + yz + zx, \quad \sqrt{\frac{x+y}{z}}, \quad \log(x^2+y-z),$$

are functions of x, y, and z.

2. *Dependent and Independent Variables.* If y is a function of x, as in the equations

$$y = x^2, \quad y = \tan 4x, \quad y = e^x,$$

x is called the *independent* variable, and y the *dependent* variable.

It is evident that whenever y is a function of x, x may be also regarded as a function of y, and the positions of dependent and independent variables reversed. Thus from the preceding equations,
$$x = \sqrt{y}, \quad x = \tfrac{1}{4}\tan^{-1}y, \quad x = \log_e y.$$

In equations involving more than two variables, as
$$z + x - y = 0, \quad w + wz + zx + y = 0,$$
one must be regarded as the dependent variable, and the others as independent variables.

3. *Explicit and Implicit Functions.* When one quantity is expressed directly in terms of another, the former is said to be an *explicit* function of the latter.

For example, y is an explicit function of x in the equations,
$$y = x^2 + 2x, \qquad y = \sqrt{x^2 + 1}.$$

When the relation between y and x is given by an equation containing these quantities, but not solved with reference to y, y is said to be an *implicit* function of x, as in the equations,
$$2xy + y^2 = x^2 + 1, \quad y + \log y = x.$$

Sometimes, as in the first of these equations, we can solve the equation with reference to y, and thus change the function from implicit to explicit. Thus we find from this equation,
$$y = -x \pm \sqrt{2x^2 + 1}.$$

4. *Algebraic and Transcendental Functions.* An *algebraic* function is one that involves only the operations of addition, subtraction, multiplication, division, involution and evolution with constant exponents. All other functions are called *transcendental* functions, including *logarithmic*, *exponential*, *trigonometric*, and *inverse trigonometric*, functions.

5. Notation of Functions. The symbols $F(x)$, $f(x)$, $\phi(x)$, $\psi(x)$, and the like, are used to denote functions of x. Thus instead of "y is a function of x," we may write

$$y = f(x) \quad \text{or} \quad y = \phi(x).$$

A functional symbol occurring more than once in the same problem or discussion is understood to denote the same function or operation, although applied to different quantities. Thus, if

$$f(x) = x^2 + 5, \quad \ldots \quad \ldots \quad \ldots \quad (1)$$

then
$$f(y) = y^2 + 5, \qquad f(a) = a^2 + 5,$$
$$f(a+1) = (a+1)^2 + 5 = a^2 + 2a + 6,$$
$$f(2) = 2^2 + 5 = 9, \qquad f(1) = 6.$$

In all these expressions $f(\ )$ denotes the same operation as defined by (1); that is, the operation of squaring the quantity and adding 5 to the result.

The following examples will further illustrate the notation of functions.

EXAMPLES.

1. If $f(x) = 2x^3 - x^2 - 7x + 6$, show that
$$f(3) = 30, \quad f(2) = 4, \quad f(0) = 6, \quad f(1) = 0,$$
$$f(-2) = 0, \quad f(\tfrac{3}{2}) = 0, \quad f(x-2) = 2x^3 - 13x^2 + 21x,$$
$$f(x+h) = 2x^3 + (6h - 1)x^2 + (6h^2 - 2h - 7)x + 2h^3$$
$$\qquad - h^2 - 7h + 6.$$

2. Given $f_1(y) = 2y^4 - y^3 + 1$, $f_2(y) = 7y^2 - 6y + 1$; show that
$$f_1(1) = f_2(1), \quad f_1(\tfrac{3}{2}) = f_2(\tfrac{3}{2}), \quad f_1(-2) = f_2(-2),$$
$$f_1(0) = f_2(0).$$

3. If $f(a) = \dfrac{a-1}{a+1}$, show that
$$\frac{f(a) - f(b)}{1 + f(a)f(b)} = \frac{a-b}{1+ab}.$$

4. If $\phi(m)=(m+1)\,m(m-1)(m-2)$, show that
$$\phi(2)=\phi(1)=\phi(0)=\phi(-1)=0,\quad \phi(3)=\phi(-2),$$
$$\frac{\phi(m+1)}{m+2}=\frac{\phi(m)}{m-2}.$$

5. If $\phi(x)=(x-a)(x-b)(x-c)$, show that
$$\phi(a)=\phi(b)=\phi(c)=0,$$
$$\frac{\phi(a+b)\cdot\phi(b+c)\cdot\phi(c+a)}{[\phi(0)]^2}=8\phi\!\left(\frac{a+b+c}{2}\right),$$
$$\frac{\phi(-a)\cdot\phi(-b)\cdot\phi(-c)}{\phi(0)}=8[\phi(a+b+c)]^2.$$

6. If $\phi(u)=e^u+e^{-u}$, show that
$$\phi(3u)=[\phi(u)]^3-3\phi(u),$$
$$\phi(u+v)\,\phi(u-v)=\phi(2u)+\phi(2v).$$

7. If $F(x)=\log\dfrac{1-x}{1+x}$, show that
$$F(x)+F(z)=F\!\left(\frac{x+z}{1+xz}\right).$$

8. If $f(x)=\log(x+\sqrt{x^2-1})$, show that
$$2f(x)=f(2x^2-1),$$
$$3f(x)=f(4x^3-3x).$$

9. Given $\psi(x)=\cos x+\sqrt{-1}\sin x$; show that
$$\psi(2a)=[\psi(a)]^2,\qquad \psi(a+b)=\psi(a)\,\psi(b).$$

10. If $f(x,y,z)=x^3+y^3+z^3-3xyz$, show that
$$f(x,y,z)f(p,q,r)=f(L,M,N),$$
where
$$L=px+qy+rz,$$
$$M=py+qz+rx,$$

CHAPTER II.

DIFFERENTIAL COEFFICIENT.

6. *Limit.* The limit of a variable quantity is a fixed value or condition, from which it can be made to differ as little as we please.

The student is supposed to be already familiar with the meaning of this term, of which the following illustrations may be mentioned.

The limit of the value of the recurring decimal $.3333\cdots$, as the number of decimal places is indefinitely increased, is $\frac{1}{3}$.

The limit of the sum of the series $1 + \frac{1}{2} + \frac{1}{4} + \frac{1}{8} + \cdots$, as the number of terms is indefinitely increased, is 2.

The circle is the limit of a regular polygon, as the number of sides is indefinitely increased.

The tangent to a curve is the limit of a secant, as the points of intersection approach coincidence.

The limit of the fraction, $\dfrac{\sin \theta}{\theta}$, as θ approaches zero, is 1, — provided θ is expressed in circular measure.

7. *Increments.* An increment of a variable quantity is any addition to its value, and is denoted by the symbol Δ written before this quantity. Thus Δx denotes an increment of x, Δy an increment of y.

For example, if we have given

$$y = x^2,$$

and assume $x = 10$, then if we increase the value of x by 2, the value of y is increased from 100 to 144, that is, by 44.

In other words, if we assume the increment of x to be $\Delta x = 2$, we shall find the increment of y to be $\Delta y = 44$.

A *negative* increment is a *decrement;* that is, a *decrease* in value.

For example, calling $x = 10$, as before, in $y = x^2$,

if $\Delta x = -2$, then $\Delta y = -36$.

8. *Differential Coefficient.* In the equation $y = x^2$, if we suppose x to vary, y will vary also. To fix the attention upon a definite value of x, let us suppose $x = 10$ and therefore $y = 100$, and let us inquire what addition or increment will be produced in y by a certain increment assigned to x. Calculating the values of Δy corresponding to different values of Δx, we find results as in the following table:

If $\Delta x =$	then $\Delta y =$	and $\frac{\Delta y}{\Delta x} =$
3.	69.	23.
2.	44.	22.
1.	21.	21.
0.1	2.01	20.1
0.01	0.2001	20.01
0.001	0.020001	20.001
h.	$20h + h^2$.	$20 + h$.

The third column gives the value of the ratio between the increments of x and of y.

It appears from the table that, as Δx diminishes and approaches zero, Δy also diminishes and approaches zero. The ratio $\frac{\Delta y}{\Delta x}$ diminishes, but instead of approaching zero, approaches 20 as its limit.

This limit of $\frac{\Delta y}{\Delta x}$ is called the *differential coefficient* of y with respect to x, and is denoted by $\frac{dy}{dx}$. In this case, when $x = 10$, $\frac{dy}{dx} = 20$.

It is to be noticed that $\frac{dy}{dx}$ is not here defined as a fraction, but as a single symbol denoting the limit of the fraction $\frac{\Delta y}{\Delta x}$. The student will find as he advances that $\frac{dy}{dx}$ has many of the properties of an ordinary fraction, and Chapter V. shows how it may be regarded as such.

9. Without restricting ourselves to any one numerical value, we may obtain $\frac{dy}{dx}$ from the equation $y = x^2$ thus:

Having $y = x^2$, let $\Delta x = h$, and let the new value of y be denoted by
$$y' = (x+h)^2;$$
therefore
$$\Delta y = y' - y = (x+h)^2 - x^2 = 2xh + h^2.$$

Dividing by $\Delta x = h$, gives
$$\frac{\Delta y}{\Delta x} = 2x + h.$$

The limit of this, when h approaches zero, is $2x$. Hence
$$\frac{dy}{dx} = 2x.$$

In the same way the differential coefficients of other given functions may be found.

For example, find $\frac{dy}{dx}$ from the equation,
$$y = 2x^3 + 1.$$
Let $\Delta x = h$,
then $y' = 2(x+h)^3 + 1.$
$$\Delta y = y' - y = 2(x+h)^3 - 2x^3 = 2(3x^2h + 3xh^2 + h^3).$$

Dividing by $\Delta x = h$ gives

$$\frac{\Delta y}{\Delta x} = 2(3x^2 + 3xh + h^2).$$

The limit of $\frac{\Delta y}{\Delta x}$ is $6x^2$, as h approaches zero.

$$\therefore \frac{dy}{dx} = 6x^2.$$

Take for another example

$$y = \sqrt{x}. \qquad \Delta x = h.$$
$$y' = \sqrt{x+h}.$$
$$\Delta y = \sqrt{x+h} - \sqrt{x}.$$
$$\frac{\Delta y}{\Delta x} = \frac{\sqrt{x+h} - \sqrt{x}}{h}.$$

The limit of this takes the indeterminate form $\frac{0}{0}$. But by rationalizing the numerator, we have

$$\frac{\Delta y}{\Delta x} = \frac{h}{h(\sqrt{x+h} + \sqrt{x})} = \frac{1}{\sqrt{x+h} + \sqrt{x}}.$$

The limit of
$$\frac{\Delta y}{\Delta x} = \frac{1}{2\sqrt{x}};$$

that is,
$$\frac{dy}{dx} = \frac{1}{2\sqrt{x}}.$$

10. *General Definition of Differential Coefficient.*

In general, if $y = \phi(x)$,

$$y' = \phi(x+h),$$
$$\Delta y = y' - y = \phi(x+h) - \phi(x),$$
$$\frac{\Delta y}{\Delta x} = \frac{\phi(x+h) - \phi(x)}{h},$$
$$\frac{dy}{dx} = \text{limit of } \frac{\phi(x+h) - \phi(x)}{h}, \text{ as } h \text{ approaches zero.}$$

The *differential coefficient* of a function may then be defined as *the limiting value of the ratio of the increment of the function to the increment of the variable, as these increments approach zero.* That is, the differential coefficient of the function $\phi(x)$ with respect to x, is

$$\text{the limit of } \frac{\phi(x+h)-\phi(x)}{h},$$

as h is indefinitely diminished.

The differential coefficient is sometimes called *the derivative*.

NOTE. — In Art. 94 will be found a geometrical illustration of the differential coefficient.

EXAMPLES.

Following the process of Art. 9, derive the following differential coefficients:

1. $y = 3x^2 - 2x.$ $\dfrac{dy}{dx} = 6x - 2.$

2. $y = x^4 + 5.$ $\dfrac{dy}{dx} = 4x^3.$

3. $y = (x-1)(2x+3).$ $\dfrac{dy}{dx} = 4x + 1.$

4. $y = \dfrac{1}{x}.$ $\dfrac{dy}{dx} = -\dfrac{1}{x^2}.$

5. $y = \dfrac{a}{x^2}.$ $\dfrac{dy}{dx} = -\dfrac{2a}{x^3}.$

6. $y = \dfrac{x-a}{x+a}.$ $\dfrac{dy}{dx} = \dfrac{2a}{(x+a)^2}.$

7. $y = x^{\frac{3}{2}}.$ $\dfrac{dy}{dx} = \dfrac{3x^{\frac{1}{2}}}{2}.$

8. $y = \sqrt{x^2 - 2}.$ $\dfrac{dy}{dx} = \dfrac{x}{\sqrt{x^2 - 2}}.$

9. $y = \dfrac{2}{\sqrt{x+1}}.$ $\dfrac{dy}{dx} = -\dfrac{1}{(x+1)^{\frac{3}{2}}}.$

10. $y = x^{\frac{1}{3}}.$ $\dfrac{dy}{dx} = \dfrac{1}{3x^{\frac{2}{3}}}.$

CHAPTER III.

DIFFERENTIATION.

11. The process of finding the differential coefficient of a given function is called *differentiation*. The examples in the preceding chapter are introduced to illustrate the meaning of the differential coefficient, but this elementary method of differentiation is too tedious for general use.

Differentiation is more readily performed by the application of certain general rules, which may be expressed by formulæ. In these formulæ u and v will denote *variable* quantities, functions of x; and c and n, *constant* quantities.

It is frequently convenient to write the differential coefficient of a quantity

$$\frac{d}{dx}u, \quad \text{instead of} \quad \frac{du}{dx}.$$

Thus the differential coefficient of $(u+v)$ is more conveniently written

$$\frac{d}{dx}(u+v), \quad \text{rather than} \quad \frac{d(u+v)}{dx}.$$

12. *Formulæ for Differentiation of Algebraic Functions.*

$$\text{I.} \quad \frac{dx}{dx} = 1.$$

$$\text{II.} \quad \frac{dc}{dx} = 0.$$

$$\text{III.} \quad \frac{d}{dx}(u+v) = \frac{du}{dx} + \frac{dv}{dx}.$$

$$\text{IV.} \quad \frac{d}{dx}(uv) = v\frac{du}{dx} + u\frac{dv}{dx}.$$

V. $\dfrac{d}{dx}(cu) = c\dfrac{du}{dx}.$

VI. $\dfrac{d}{dx}\left(\dfrac{u}{v}\right) = \dfrac{v\dfrac{du}{dx} - u\dfrac{dv}{dx}}{v^2}.$

VII. $\dfrac{d}{dx}(u^n) = nu^{n-1}\dfrac{du}{dx}.$

These formulæ express the following general rules of differentiation:

I. *The differential coefficient of a variable with respect to itself is unity.*

II. *The differential coefficient of a constant is zero.*

III. *The differential coefficient of the sum of two variables is the sum of their differential coefficients.*

IV. *The differential coefficient of the product of two variables is the sum of the products of each variable by the differential coefficient of the other.*

V. *The differential coefficient of the product of a constant and a variable is the product of the constant and the differential coefficient of the variable.*

VI. *The differential coefficient of a fraction is the differential coefficient of the numerator multiplied by the denominator minus the differential coefficient of the denominator multiplied by the numerator, this difference being divided by the square of the denominator.*

VII. *The differential coefficient of any power of a variable is the product of the exponent, the power with exponent diminished by 1, and the differential coefficient of the variable.*

13. *Derivation of Formulæ.*

Proof of I. This follows immediately from the definition of a differential coefficient. For since $\dfrac{\Delta x}{\Delta x} = 1$, its limit $\dfrac{dx}{dx} = 1.$

Proof of II. A constant is a quantity whose value does not vary. Hence

$$\Delta c = 0 \quad \text{and} \quad \frac{\Delta c}{\Delta x} = 0;$$

therefore its limit $\quad \dfrac{dc}{dx} = 0.$

Proof of III. Let $y = u + v$, and suppose that when x is changed into $x + h$, y, u, and v become y', u', and v'; then

$$y' = u' + v';$$

therefore $\quad y' - y = u' - u + v' - v;$

that is, $\quad \Delta y = \Delta u + \Delta v.$

Divide by Δx; then

$$\frac{\Delta y}{\Delta x} = \frac{\Delta u}{\Delta x} + \frac{\Delta v}{\Delta x}.$$

Now suppose Δx to diminish and approach zero, and we have, for the limits of these fractions,

$$\frac{dy}{dx} = \frac{du}{dx} + \frac{dv}{dx}.$$

If in this we substitute for y, $u + v$, we have

$$\frac{d}{dx}(u + v) = \frac{du}{dx} + \frac{dv}{dx}.$$

It is evident that the same proof would apply to any number of variables connected by plus or minus signs. We should then have

$$\frac{d}{dx}(u \pm v \pm w \pm \cdots) = \frac{du}{dx} \pm \frac{dv}{dx} \pm \frac{dw}{dx} \pm \cdots.$$

Proof of IV. Let $y = uv$; then

$$y' = u'v',$$

and $\quad y' - y = u'v' - uv = (u' - u)v' + u(v' - v);$

that is, $\quad \Delta y = v'\Delta u + u\Delta v.$

Divide by Δx; then

$$\frac{\Delta y}{\Delta x} = v'\frac{\Delta u}{\Delta x} + u\frac{\Delta v}{\Delta x}.$$

Now suppose Δx to approach zero, and, noticing that the limit of v' is v, we have

$$\frac{dy}{dx} = v\frac{du}{dx} + u\frac{dv}{dx};$$

that is,
$$\frac{d}{dx}(uv) = v\frac{du}{dx} + u\frac{dv}{dx}.$$

This formula may be extended to the product of three or more variables. Thus we have

$$\frac{d}{dx}(uvw) = \frac{d}{dx}(uv \cdot w) = w\frac{d}{dx}(uv) + uv\frac{dw}{dx}$$

$$= w\left(v\frac{du}{dx} + u\frac{dv}{dx}\right) + uv\frac{dw}{dx}$$

$$= vw\frac{du}{dx} + uw\frac{dv}{dx} + uv\frac{dw}{dx}.$$

Similarly, for the product of four functions, we have

$$\frac{d}{dx}(uvwz) = vwz\frac{du}{dx} + wzu\frac{dv}{dx} + zuv\frac{dw}{dx} + uvw\frac{dz}{dx}.$$

A similar relation holds for the product of any number of variables.

Proof of V. This is a special case of IV., $\frac{dc}{dx}$ being zero. But we may derive it independently thus:

$$y = cu,$$
$$y' = cu',$$
$$y' - y = c(u' - u),$$
$$\Delta y = c\Delta u,$$
$$\frac{\Delta y}{\Delta x} = c\frac{\Delta u}{\Delta x},$$

$$\frac{dy}{dx} = c\frac{du}{dx}, \quad \text{or} \quad \frac{d}{dx}(cu) = c\frac{du}{dx}.$$

Proof of VI. Let
$$y = \frac{u}{v},$$

then
$$y' = \frac{u'}{v'};$$

therefore $\quad y' - y = \dfrac{u'}{v'} - \dfrac{u}{v} = \dfrac{u'v - uv'}{v'v} = \dfrac{(u'-u)v - u(v'-v)}{v'v};$

that is,
$$\Delta y = \frac{v\Delta u - u\Delta v}{v'v},$$

$$\frac{\Delta y}{\Delta x} = \frac{v\dfrac{\Delta u}{\Delta x} - u\dfrac{\Delta v}{\Delta x}}{v'v}.$$

Now suppose Δx to diminish towards zero, and, noticing that the limit of v' is v, we have

$$\frac{dy}{dx} = \frac{v\dfrac{du}{dx} - u\dfrac{dv}{dx}}{v^2}.$$

Or we may derive VI. from IV. thus:

Since
$$y = \frac{u}{v},$$

therefore $\quad yv = u.$

By IV.,
$$v\frac{dy}{dx} + y\frac{dv}{dx} = \frac{du}{dx},$$

$$v\frac{dy}{dx} = \frac{du}{dx} - \frac{u}{v}\frac{dv}{dx};$$

therefore
$$\frac{dy}{dx} = \frac{v\dfrac{du}{dx} - u\dfrac{dv}{dx}}{v^2}.$$

DIFFERENTIATION.

Proof of VII. *First*, suppose n to be a positive integer.

Let $\quad y = u^n$,

and, $\quad y' = u'^n$,

$$y' - y = u'^n - u^n$$
$$= (u' - u)(u'^{n-1} + u'^{n-2}u + u'^{n-3}u^2 \cdots + u^{n-1});$$

that is, $\quad \Delta y = \Delta u (u'^{n-1} + u'^{n-2}u + u'^{n-3}u^2 \cdots + u^{n-1})$,

$$\frac{\Delta y}{\Delta x} = (u'^{n-1} + u'^{n-2}u + u'^{n-3}u^2 \cdots + u^{n-1})\frac{\Delta u}{\Delta x}.$$

Now let Δx diminish; then, u being the limit of u', each of the n terms within the parenthesis becomes u^{n-1}; therefore

$$\frac{dy}{dx} = nu^{n-1}\frac{du}{dx}.$$

Second, suppose n to be a positive fraction, $\frac{p}{q}$.

Let $\quad y = u^{\frac{p}{q}}$,

then $\quad y^q = u^p$;

therefore $\quad \dfrac{d}{dx}(y^q) = \dfrac{d}{dx}(u^p).$

But we have already shown VII. to be true when the exponent is a positive integer; hence we may apply it to each member of this equation. This gives

$$qy^{q-1}\frac{dy}{dx} = pu^{p-1}\frac{du}{dx};$$

therefore $\quad \dfrac{dy}{dx} = \dfrac{p}{q}\dfrac{u^{p-1}}{y^{q-1}}\dfrac{du}{dx}.$

Substituting for y, $u^{\frac{p}{q}}$, gives

$$\frac{dy}{dx} = \frac{p}{q}\frac{u^{p-1}}{u^{p-\frac{p}{q}}}\frac{du}{dx} = \frac{p}{q}u^{\frac{p}{q}-1}\frac{du}{dx},$$

which shows VII. to be true in this case also. Hence that formula applies to any positive value of n, whether integral or fractional.

Third, suppose n to be negative and equal to $-m$.

Let $$y = u^{-m} = \frac{1}{u^m};$$

by VI., $$\frac{dy}{dx} = \frac{-\frac{d}{dx}(u^m)}{u^{2m}} = \frac{-mu^{m-1}\frac{du}{dx}}{u^{2m}} = -mu^{-m-1}\frac{du}{dx}.$$

Hence VII. is universally true.

EXAMPLES.

Differentiate the following functions:

1. $y = x^4$.

If two quantities are equal, their differential coefficients must be equal. Hence
$$\frac{dy}{dx} = \frac{d}{dx}(x^4).$$

If we apply VII., substituting $u = x$ and $n = 4$, we have
$$\frac{d}{dx}(x^4) = 4x^3\frac{dx}{dx} = 4x^3, \quad \text{by I.}$$

$$\therefore \frac{dy}{dx} = 4x^3.$$

2. $y = 3x^4 + 4x^3$.

$$\frac{dy}{dx} = \frac{d}{dx}(3x^4 + 4x^3) = \frac{d}{dx}(3x^4) + \frac{d}{dx}(4x^3),$$

by III., making $u = 3x^4$ and $v = 4x^3$.

$$\frac{d}{dx}(3x^4) = 3\frac{d}{dx}(x^4), \quad \text{by V.,}$$
$$= 3 \cdot 4x^3 = 12x^3.$$

Similarly, $\frac{d}{dx}(4x^3) = 4\frac{d}{dx}(x^3) = 4 \cdot 3x^2 = 12x^2$.

$$\therefore \frac{dy}{dx} = 12x^3 + 12x^2 = 12(x^3 + x^2).$$

3. $y = x^{\frac{3}{2}} + 2.$

$$\frac{dy}{dx} = \frac{d}{dx}(x^{\frac{3}{2}}) + \frac{d}{dx}(2).$$

$$\frac{d}{dx}(x^{\frac{3}{2}}) = \frac{3}{2}x^{\frac{1}{2}}, \quad \text{by VII.}$$

$$\frac{d}{dx}(2) = 0, \quad \text{by II.}$$

$$\therefore \frac{dy}{dx} = \frac{3}{2}x^{\frac{1}{2}}.$$

4. $y = 3\sqrt{x} - \dfrac{2}{\sqrt{x}} + \dfrac{1}{x^3} + a.$

$$\frac{dy}{dx} = \frac{d}{dx}(3x^{\frac{1}{2}}) - \frac{d}{dx}(2x^{-\frac{1}{2}}) + \frac{d}{dx}(x^{-3}) + \frac{da}{dx}$$

$$= \frac{3}{2}x^{-\frac{1}{2}} - 2\left(-\frac{1}{2}\right)x^{-\frac{3}{2}} - 3x^{-4} + 0$$

$$= \frac{3}{2x^{\frac{1}{2}}} + \frac{1}{x^{\frac{3}{2}}} - \frac{3}{x^4}.$$

5. $y = \dfrac{x+3}{x^2+3}.$

$$\frac{dy}{dx} = \frac{d}{dx}\left(\frac{x+3}{x^2+3}\right).$$

Applying VI., making

$$u = x + 3 \text{ and } v = x^2 + 3, \text{ we have}$$

$$\frac{d}{dx}\left(\frac{x+3}{x^2+3}\right) = \frac{(x^2+3)\frac{d}{dx}(x+3) - (x+3)\frac{d}{dx}(x^2+3)}{(x^2+3)^2}$$

$$= \frac{x^2 + 3 - (x+3)2x}{(x^2+3)^2} = \frac{3 - 6x - x^2}{(x^2+3)^2}.$$

$$\therefore \frac{dy}{dx} = \frac{3 - 6x - x^2}{(x^2+3)^2}.$$

6. $y = (x^2+2)^{\frac{2}{3}}.$

$$\frac{dy}{dx} = \frac{d}{dx}(x^2+2)^{\frac{2}{3}}.$$

If we apply VII., making

$$u = x^2+2 \quad \text{and} \quad n = \frac{2}{3}, \quad \text{we have}$$

$$\frac{d}{dx}(x^2+2)^{\frac{2}{3}} = \frac{2}{3}(x^2+2)^{-\frac{1}{3}}\frac{d}{dx}(x^2+2)$$

$$= \frac{2}{3}(x^2+2)^{-\frac{1}{3}}2x = \frac{4x}{3(x^2+2)^{\frac{1}{3}}}.$$

$$\therefore \frac{dy}{dx} = \frac{4x}{3(x^2+2)^{\frac{1}{3}}}.$$

7. $y = (x^2+1)\sqrt{x^3-x}.$

$$\frac{dy}{dx} = \frac{d}{dx}[(x^2+1)(x^3-x)^{\frac{1}{2}}].$$

If we apply IV., making

$$u = x^2+1 \quad \text{and} \quad v = (x^3-x)^{\frac{1}{2}}, \quad \text{we have}$$

$$\frac{d}{dx}[(x^2+1)(x^3-x)^{\frac{1}{2}}]$$

$$= (x^2+1)\frac{d}{dx}(x^3-x)^{\frac{1}{2}} + (x^3-x)^{\frac{1}{2}}\frac{d}{dx}(x^2+1).$$

$$\frac{d}{dx}(x^3-x)^{\frac{1}{2}} = \frac{1}{2}(x^3-x)^{-\frac{1}{2}}\frac{d}{dx}(x^3-x) = \frac{1}{2}(x^3-x)^{-\frac{1}{2}}(3x^2-1).$$

$$\frac{d}{dx}(x^2+1) = 2x.$$

$$\therefore \frac{dy}{dx} = \frac{1}{2}(x^2+1)(3x^2-1)(x^3-x)^{-\frac{1}{2}} + (x^3-x)^{\frac{1}{2}}2x$$

$$= \frac{(x^2+1)(3x^2-1) + 4x(x^3-x)}{2(x^3-x)^{\frac{1}{2}}} = \frac{7x^4-2x^2-1}{2(x^3-x)^{\frac{1}{2}}}.$$

DIFFERENTIATION.

8. $y = (x+1)^5(2x-1)^3.$ $\quad \dfrac{dy}{dx} = (16x+1)(x+1)^4(2x-1)^2.$

9. $y = \dfrac{a+bx+cx^2}{x}.$ $\quad \dfrac{dy}{dx} = c - \dfrac{a}{x^2}.$

10. $y = \dfrac{(x-1)^3}{x^{\frac{1}{3}}}.$ $\quad \dfrac{dy}{dx} = \dfrac{8}{3}x^{\frac{5}{3}} - 5x^{\frac{2}{3}} + 2x^{-\frac{1}{3}} + \dfrac{1}{3}x^{-\frac{4}{3}}.$

11. $y = \dfrac{x^{\frac{3}{2}}+x-x^{\frac{1}{2}}+a}{x^{\frac{1}{2}}}.$ $\quad \dfrac{dy}{dx} = \dfrac{2x^{\frac{3}{2}}-x+2x^{\frac{1}{2}}-3a}{2x^{\frac{3}{2}}}.$

12. Given

$$(a+x)^5 = a^5 + 5a^4x + 10a^3x^2 + 10a^2x^3 + 5ax^4 + x^5;$$

derive by differentiation the expansion of $(a+x)^4$.

13. Given $1 + x + x^2 \cdots + x^n = \dfrac{x^{n+1}-1}{x-1};$

derive the sum of the series $1 + 2x + 3x^2 \cdots + nx^{n-1}.$

$$\text{Ans. } \dfrac{nx^{n+1}-(n+1)x^n+1}{(x-1)^2}.$$

14. $y = \sqrt{\dfrac{1+x}{1-x}}.$ $\quad \dfrac{dy}{dx} = \dfrac{1}{(1-x)\sqrt{1-x^2}}.$

15. $y = \dfrac{x^n}{(1+x)^n}.$ $\quad \dfrac{dy}{dx} = \dfrac{nx^{n-1}}{(1+x)^{n+1}}.$

16. $y = (1 - 2x + 3x^2 - 4x^3)(1+x)^2.$ $\quad \dfrac{dy}{dx} = -20x^3(1+x).$

17. $y = (1 - 3x^2 + 6x^4)(1+x^2)^3.$ $\quad \dfrac{dy}{dx} = 60x^5(1+x^2)^2.$

18. $y = x^5 (a+3x)^3 (a-2x)^2$.

$$\frac{dy}{dx} = 5x^4(a+3x)^2(a-2x)(a^2+2ax-12x^2)$$

19. $y = x^{15}(a-3x)^5(a+5x)^3$.

$$\frac{dy}{dx} = 15x^{14}(a-3x)^4(a+5x)^2(a^2+2ax-23x^2).$$

20. $y = (a+x)^m (b+x)^n$.

$$\frac{dy}{dx} = [m(b+x) + n(a+x)](a+x)^{m-1}(b+x)^{n-1}$$

21. $y = \dfrac{1}{(a+x)^m(b+x)^n}$.

$$\frac{dy}{dx} = -\frac{m(b+x) + n(a+x)}{(a+x)^{m+1}(b+x)^{n+1}}.$$

22. $y = \dfrac{x}{\sqrt{1-x^2}}$. $\qquad \dfrac{dy}{dx} = \dfrac{1}{(1-x^2)^{\frac{3}{2}}}$.

23. $y = \dfrac{1-x}{\sqrt{1+x^2}}$. $\qquad \dfrac{dy}{dx} = -\dfrac{1+x}{(1+x^2)^{\frac{3}{2}}}$.

24. $y = \dfrac{2\sqrt{x}}{3+x^2}$. $\qquad \dfrac{dy}{dx} = \dfrac{3(1-x^2)}{(3+x^2)^2\sqrt{x}}$.

25. $y = \dfrac{1}{x+\sqrt{1+x^2}}$. $\qquad \dfrac{dy}{dx} = \dfrac{x}{\sqrt{1+x^2}} - 1$.

26. $y = \dfrac{x}{x+\sqrt{1-x^2}}$. $\qquad \dfrac{dy}{dx} = \dfrac{1}{2x(1-x^2)+\sqrt{1-x^2}}$.

27. $y = \dfrac{\sqrt{a+x}+\sqrt{a-x}}{\sqrt{a+x}-\sqrt{a-x}}$. $\qquad \dfrac{dy}{dx} = -\dfrac{a^2+a\sqrt{a^2-x^2}}{x^2\sqrt{a^2-x^2}}$.

28. $y = \dfrac{3x^2+2}{x(x^2+1)^{\frac{2}{3}}}$. $\qquad \dfrac{dy}{dx} = -\dfrac{2}{x^2(x^2+1)^{\frac{4}{3}}}$.

29. $y = 3(x^2+1)^{\frac{4}{3}}(4x^2-3)$. $\qquad \dfrac{dy}{dx} = 56x^3(x^2+1)^{\frac{1}{3}}$.

DIFFERENTIATION. 21

30. $y = \dfrac{\sqrt{(x+a)^3}}{\sqrt{x-a}}.$ $\qquad \dfrac{dy}{dx} = \dfrac{(x-2a)\sqrt{x+a}}{(x-a)^{\frac{3}{2}}}.$

31. $y = \dfrac{\sqrt{1+x^2}+\sqrt{1-x^2}}{\sqrt{1+x^2}-\sqrt{1-x^2}}.$ $\qquad \dfrac{dy}{dx} = -\dfrac{2}{x^3}\left(1 + \dfrac{1}{\sqrt{1-x^4}}\right).$

32. $y = \left(\dfrac{x}{1+\sqrt{1-x^2}}\right)^n.$ $\qquad \dfrac{dy}{dx} = \dfrac{ny}{x\sqrt{1-x^2}}.$

14. Formulæ for Differentiation of Logarithmic and Exponential Functions.

$\qquad$ VIII. $\quad \dfrac{d}{dx}\log_a u = \log_a e \dfrac{\frac{du}{dx}}{u}.$

$\qquad$ IX. $\quad \dfrac{d}{dx}\log_e u = \dfrac{\frac{du}{dx}}{u}.$

$\qquad$ X. $\quad \dfrac{d}{dx}a^u = \log_e a \cdot a^u \dfrac{du}{dx}.$

$\qquad$ XI. $\quad \dfrac{d}{dx}e^u = e^u \dfrac{du}{dx}.$

$\qquad$ XII. $\quad \dfrac{d}{dx}u^v = vu^{v-1}\dfrac{du}{dx} + \log_e u \cdot u^v \dfrac{dv}{dx}.$

15. Before deriving these formulæ it is necessary to find the limit of the expression

$$\left(1 + \frac{1}{z}\right)^z, \text{ as } z \text{ approaches infinity.}$$

By the Binomial Theorem

$$\left(1+\frac{1}{z}\right)^z = 1 + z\cdot\frac{1}{z} + \frac{z(z-1)}{\underline{|2}}\left(\frac{1}{z}\right)^2 + \frac{z(z-1)(z-2)}{\underline{|3}}\left(\frac{1}{z}\right)^3 + \cdots,$$

which may be written

$$\left(1+\frac{1}{z}\right)^z = 1 + 1 + \frac{1-\frac{1}{z}}{\underline{|2}} + \frac{\left(1-\frac{1}{z}\right)\left(1-\frac{2}{z}\right)}{\underline{|3}} + \cdots.$$

Now when z increases indefinitely, we have
$$\text{limit of } \left(1+\frac{1}{z}\right)^z = 1 + 1 + \frac{1}{\lfloor 2} + \frac{1}{\lfloor 3} + \cdots.$$
This quantity is usually denoted by e, so that
$$e = 1 + \frac{1}{1} + \frac{1}{\lfloor 2} + \frac{1}{\lfloor 3} + \cdots.$$

The value of e can be easily calculated to any desired number of decimals by computing the values of the successive terms of this series. For seven decimal places the calculation is as follows,—

```
          1.
          1.
           .5
           .166666667
           .041666667
           .008333333
           .001388889
           .000198413
           .000024802
           .000002756
           .000000276
           .000000025
           .000000002
       ─────────────
   e =  2.7182818 ···
```

By calculating the value of $\left(1+\frac{1}{z}\right)^z$ for different values of z, we may verify its limit. Thus

$$(1 + \tfrac{1}{2})^2 = 2.25$$
$$(1 + \tfrac{1}{5})^5 = 2.48832$$
$$(1 + \tfrac{1}{10})^{10} = 2.59374$$
$$(1.01)^{100} = 2.70481$$
$$(1.001)^{1000} = 2.71692$$
$$(1.0001)^{10000} = 2.71815$$
$$(1.00001)^{100000} = 2.71827$$
$$(1.000001)^{1000000} = 2.71828$$

16. Derivation of Formulæ.

Proof of VIII. Let $y = \log_a u$,

then
$$y' = \log_a(u + \Delta u),$$
$$\Delta y = \log_a(u + \Delta u) - \log_a u = \log_a \frac{u + \Delta u}{u}$$
$$= \log_a \left(1 + \frac{\Delta u}{u}\right) = \frac{\Delta u}{u} \log_a \left(1 + \frac{\Delta u}{u}\right)^{\frac{u}{\Delta u}}.$$

Dividing by Δx,
$$\frac{\Delta y}{\Delta x} = \log_a \left(1 + \frac{\Delta u}{u}\right)^{\frac{u}{\Delta u}} \frac{\Delta u}{\Delta x}\cdot\frac{1}{u}.$$

Now if Δx approach zero, Δu at the same time approaches zero; then the limit of $\left(1 + \frac{\Delta u}{u}\right)^{\frac{u}{\Delta u}}$ is the same as the limit of $\left(1 + \frac{1}{z}\right)^z$ as z increases indefinitely. But in Art. 15 we have already found the latter limit to be e. Hence we have

$$\frac{dy}{dx} = \log_a e \frac{\frac{du}{dx}}{u}.$$

Proof of IX. This is a special case of VIII., when $a = e$. In this case
$$\log_a e = \log_e e = 1.$$

NOTE. — Logarithms to base e are called *Napierian* logarithms. Hereafter, when no base is specified, Napierian logarithms are to be understood.

That is
$$\log u = \log_e u.$$

Proof of X.

Let
$$y = a^u.$$

Taking the logarithm of each member, we have
$$\log y = u \log a;$$

therefore by IX.,
$$\frac{\frac{dy}{dx}}{y} = \log a \frac{du}{dx}.$$

Multiplying by $y = a^u$, we have

$$\frac{dy}{dx} = \log a \cdot a^u \frac{du}{dx}.$$

Proof of XI. This is a special case of X., where $a = e$.

Proof of XII. Let $y = u^v$.

Taking the logarithm of each member, we have

$$\log y = v \log u;$$

therefore by IX.,
$$\frac{\frac{dy}{dx}}{y} = \frac{v \frac{du}{dx}}{u} + \log u \frac{dv}{dx}.$$

Multiplying by $y = u^v$, we have

$$\frac{dy}{dx} = vu^{v-1}\frac{du}{dx} + \log u \cdot u^v \frac{dv}{dx}.$$

EXAMPLES.

1. $y = \log(3x^2 + x)$. $\qquad \dfrac{dy}{dx} = \dfrac{6x+1}{3x^2+x}.$

2. $y = x \log x$. $\qquad \dfrac{dy}{dx} = 1 + \log x.$

3. $y = x^n \log x$. $\qquad \dfrac{dy}{dx} = x^{n-1}(1 + n \log x)$

4. $y = \log \sqrt{1 - x^2}$. $\qquad \dfrac{dy}{dx} = -\dfrac{x}{1 - x^2}.$

5. $y = e^x(1 - x^3)$. $\qquad \dfrac{dy}{dx} = e^x(1 - 3x^2 - x^3).$

6. $y = \sqrt{x} - \log(\sqrt{x} + 1)$. $\qquad \dfrac{dy}{dx} = \dfrac{1}{2(\sqrt{x}+1)}.$

7. $y = \log(\log x)$. $\qquad \dfrac{dy}{dx} = \dfrac{1}{x \log x}.$

8. $y = \log(e^x + e^{-x})$. $\qquad \dfrac{dy}{dx} = \dfrac{e^{2x} - 1}{e^{2x} + 1}.$

9. $y = (x-3)e^{2x} + 4xe^x + x.$ $\dfrac{dy}{dx} = (2x-5)e^{2x} + 4(x+1)e^x + 1.$

10. $y = \log_{10}(5x + x^3).$ $\dfrac{dy}{dx} = M\dfrac{5 + 3x^2}{5x + x^3},$

where $M = \dfrac{1}{\log_e 10} = \log_{10} e = .434294$

11. $y = 5^{x^2 + 2x}.$ $\dfrac{dy}{dx} = 2(x+1)5^{x^2+2x}\log 5,$

$\log 5 = 1.609440$

12. $y = \dfrac{e^x - e^{-x}}{e^x + e^{-x}}.$ $\dfrac{dy}{dx} = \dfrac{4}{(e^x + e^{-x})^2}.$

What is the result of differentiating both members of each of the three following equations?

13. $\log(1 + x) = x - \dfrac{x^2}{2} + \dfrac{x^3}{3} - \dfrac{x^4}{4} + \cdots$

Ans. $\dfrac{1}{1+x} = 1 - x + x^2 - x^3 + \cdots$

14. $\log\dfrac{1+x}{1-x} = 2\left(x + \dfrac{x^3}{3} + \dfrac{x^5}{5} + \dfrac{x^7}{7} + \cdots\right).$

Ans. $\dfrac{1}{1-x^2} = 1 + x^2 + x^4 + x^6 + \cdots$

15. $e^x = 1 + x + \dfrac{x^2}{\lfloor 2} + \dfrac{x^3}{\lfloor 3} + \dfrac{x^4}{\lfloor 4} + \cdots$

Ans. $e^x = 1 + x + \dfrac{x^2}{\lfloor 2} + \dfrac{x^3}{\lfloor 3} + \cdots$

16. $y = x^n a^x.$ $\dfrac{dy}{dx} = x^{n-1}a^x(n + x\log a).$

17. $y = \log(x-2) - \dfrac{4(x-1)}{(x-2)^2}.$ $\dfrac{dy}{dx} = \dfrac{x^3 + 4}{(x-2)^3}.$

18. $y = \log\dfrac{\sqrt{a} + \sqrt{x}}{\sqrt{a} - \sqrt{x}}.$ $\dfrac{dy}{dx} = \dfrac{\sqrt{a}}{(a-x)\sqrt{x}}.$

19. $y = \dfrac{x \log x}{1-x} + \log(1-x).$ $\dfrac{dy}{dx} = \dfrac{\log x}{(1-x)^2}.$

20. $y = e^{\sqrt{x}}(x^{\frac{5}{2}} - 3x + 6x^{\frac{1}{2}} - 6).$ $\dfrac{dy}{dx} = \dfrac{1}{2} x e^{\sqrt{x}}.$

21. $y = \dfrac{x^4}{4}\left[(\log x)^2 - \log \sqrt{x} + \frac{1}{8}\right].$ $\dfrac{dy}{dx} = x^3 (\log x)^2.$

22. $y = e^{ax}\left(x^3 - \dfrac{3x^2}{a} + \dfrac{6x}{a^2} - \dfrac{6}{a^3}\right).$ $\dfrac{dy}{dx} = ax^3 e^{ax}.$

23. $y = \log x \cdot \log(\log x) - \log x.$ $\dfrac{dy}{dx} = \dfrac{\log(\log x)}{x}.$

24. $y = \log(x - 3 + \sqrt{x^2 - 6x + 13}).$ $\dfrac{dy}{dx} = \dfrac{1}{\sqrt{x^2 - 6x + 13}}.$

25. $y = m \log(\sqrt{x} + \sqrt{x+m}) + \sqrt{mx + x^2}.$
$$\dfrac{dy}{dx} = \sqrt{\dfrac{m+x}{x}}.$$

26. $y = \log \dfrac{x}{a - \sqrt{a^2 - x^2}}.$ $\dfrac{dy}{dx} = -\dfrac{a}{x\sqrt{a^2 - x^2}}.$

27. $y = \log \dfrac{x\sqrt{2} + \sqrt{1 + x^4}}{\sqrt{1 - x^4}}.$ $\dfrac{dy}{dx} = \dfrac{\sqrt{2}}{(1 - x^4)\sqrt{1 + x^4}}.$

28. $y = \log \dfrac{\sqrt{x^2 + a^2} + \sqrt{x^2 + b^2}}{\sqrt{x^2 + a^2} - \sqrt{x^2 + b^2}}.$ $\dfrac{dy}{dx} = \dfrac{2x}{\sqrt{x^2 + a^2}\sqrt{x^2 + b^2}}.$

29. $y = \log \sqrt{\dfrac{x-1}{x+1}} + \log \sqrt{\dfrac{x^2 + 1}{x^2 - 1}}.$ $\dfrac{dy}{dx} = \dfrac{x^3 - 1}{x^4 + x^2 + 1}.$

30. $y = (e^x - e^{-x})^2 (e^{2x} + 2e^{4x} + 3e^{6x}).$ $\dfrac{dy}{dx} = 24 e^{6x}(e^{2x} - 1).$

31. $y = x^{\frac{1}{x}}.$ $\dfrac{dy}{dx} = x^{\frac{1-2x}{x}}(1 - \log x).$

32. $y = \left(\dfrac{x}{n}\right)^{nx}.$ $\dfrac{dy}{dx} = n\left(\dfrac{x}{n}\right)^{nx}\left(1 + \log \dfrac{x}{n}\right).$

33. $y = (ex)^x$. $\quad \dfrac{dy}{dx} = (ex)^x(2 + \log x)$.

34. $y = \left(\dfrac{x}{e}\right)^{\frac{x}{e}}$. $\quad \dfrac{dy}{dx} = \dfrac{1}{e}\left(\dfrac{x}{e}\right)^{\frac{x}{e}} \log x$.

35. $y = x^{\log x}$. $\quad \dfrac{dy}{dx} = \log x^2 \cdot x^{\log x - 1}$.

36. $y = x^{\frac{1}{\log x}}$. $\quad \dfrac{dy}{dx} = 0$.

37. $y = e^{e^x}$. $\quad \dfrac{dy}{dx} = e^{e^x} e^x$.

38. $y = e^{x^x}$. $\quad \dfrac{dy}{dx} = e^{x^x} x^x (1 + \log x)$.

39. $y = x^{x^x}$. $\quad \dfrac{dy}{dx} = y x^x \left[\dfrac{1}{x} + \log x + (\log x)^2\right]$.

17. *Formulæ for Differentiation of Trigonometric Functions.* In the following formulæ the angle u is supposed to be expressed in circular measure.

$$\text{XIII.} \quad \frac{d}{dx}\sin u = \cos u \frac{du}{dx}.$$

$$\text{XIV.} \quad \frac{d}{dx}\cos u = -\sin u \frac{du}{dx}.$$

$$\text{XV.} \quad \frac{d}{dx}\tan u = \sec^2 u \frac{du}{dx}.$$

$$\text{XVI.} \quad \frac{d}{dx}\cot u = -\csc^2 u \frac{du}{dx}.$$

$$\text{XVII.} \quad \frac{d}{dx}\sec u = \sec u \tan u \frac{du}{dx}.$$

$$\text{XVIII.} \quad \frac{d}{dx}\csc u = -\csc u \cot u \frac{du}{dx}.$$

$$\text{XIX.} \quad \frac{d}{dx}\operatorname{vers} u = \sin u \frac{du}{dx}.$$

18. Derivation of Formulæ.

Proof of XIII. Let $y = \sin u$,

then $\qquad y' = \sin(u + \Delta u);$

therefore $\qquad \Delta y = \sin(u + \Delta u) - \sin u.$

But from Trigonometry,

$$\sin A - \sin B = 2 \sin \tfrac{1}{2}(A - B) \cos \tfrac{1}{2}(A + B).$$

If we substitute $A = u + \Delta u$ and $B = u,$

we have $\qquad \Delta y = 2 \cos\left(u + \dfrac{\Delta u}{2}\right) \sin \dfrac{\Delta u}{2}.$

Hence $\qquad \dfrac{\Delta y}{\Delta x} = \cos\left(u + \dfrac{\Delta u}{2}\right) \dfrac{\sin \dfrac{\Delta u}{2}}{\dfrac{\Delta u}{2}} \dfrac{\Delta u}{\Delta x}.$

Now when Δx approaches zero, Δu likewise approaches zero, and as Δu is in circular measure, the limit of

$$\dfrac{\sin \dfrac{\Delta u}{2}}{\dfrac{\Delta u}{2}} \text{ is unity.}$$

Hence $\qquad \dfrac{dy}{dx} = \cos u \dfrac{du}{dx}.$

Proof of XIV. This may be derived by substituting in XIII. for u, $\dfrac{\pi}{2} - u.$

Then $\qquad \dfrac{d}{dx} \sin\left(\dfrac{\pi}{2} - u\right) = \cos\left(\dfrac{\pi}{2} - u\right) \dfrac{d}{dx}\left(\dfrac{\pi}{2} - u\right),$

or
$$\frac{d}{dx}\cos u = \sin u\left(-\frac{du}{dx}\right) = -\sin u\frac{du}{dx}.$$

Proof of XV. Since $\tan u = \dfrac{\sin u}{\cos u}$,

by VI.,
$$\frac{d}{dx}\tan u = \frac{\cos u\dfrac{d}{dx}\sin u - \sin u\dfrac{d}{dx}\cos u}{\cos^2 u}$$

$$= \frac{\cos^2 u\dfrac{du}{dx} + \sin^2 u\dfrac{du}{dx}}{\cos^2 u} = \frac{\dfrac{du}{dx}}{\cos^2 u}$$

$$= \sec^2 u\frac{du}{dx}.$$

Proof of XVI. This may be derived from XV. by substituting $\frac{\pi}{2} - u$ for u.

Proof of XVII. Since $\sec u = \dfrac{1}{\cos u}$,

by VI.,
$$\frac{d}{dx}\sec u = \frac{-\dfrac{d}{dx}\cos u}{\cos^2 u} = \frac{\sin u\dfrac{du}{dx}}{\cos^2 u}$$

$$= \sec u \tan u \frac{du}{dx}.$$

Proof of XVIII. This may be derived from XVII. by substituting $\frac{\pi}{2} - u$ for u.

Proof of XIX. This is readily obtained from XIV. by the relation
$$\operatorname{vers} u = 1 - \cos u.$$

EXAMPLES.

1. $y = \sin 2x \cos x.$ $\quad \dfrac{dy}{dx} = 2\cos 2x \cos x - \sin 2x \sin x$

2. $y = \tan^2 5x.$ $\quad \dfrac{dy}{dx} = 10 \tan 5x \sec^2 5x.$

3. $y = \tan x - x.$ $\quad \dfrac{dy}{dx} = \tan^2 x.$

4. $y = \sin(nx + m).$ $\quad \dfrac{dy}{dx} = n \cos(nx + m).$

5. $y = \dfrac{\tan x - 1}{\sec x}.$ $\quad \dfrac{dy}{dx} = \sin x + \cos x.$

6. $y = \sin^3 x \cos x.$ $\quad \dfrac{dy}{dx} = \sin^2 x (3 \cos^2 x - \sin^2 x).$

7. $y = \sin(x+a) \cos(x-a). \quad \dfrac{dy}{dx} = \cos 2x.$

8. $y = \dfrac{\sin(a-x)}{\sin(a+x)}.$ $\quad \dfrac{dy}{dx} = -\sin 2a \, \csc^2(a+x).$

9. $y = \tan^2 x - \log(\sec^2 x).$ $\quad \dfrac{dy}{dx} = 2 \tan^3 x.$

10. $y = \tan^4 x - 2 \tan^2 x + \log(\sec^4 x).$

    $\dfrac{dy}{dx} = 4 \tan^5 x.$

11. $y = (a \sin^2 x + b \cos^2 x)^n.$

    $\dfrac{dy}{dx} = n(a-b) \sin 2x (a \sin^2 x + b \cos^2 x)^{n-1}.$

12. $y = \log \sin x.$ $\quad \dfrac{dy}{dx} = \cot x.$

13. $y = \log \tan x.$ $\quad \dfrac{dy}{dx} = \dfrac{2}{\sin 2x}.$

14. $y = \log \sec x.$ $\quad \dfrac{dy}{dx} = \tan x.$

15. $y = \text{vers}\left(\dfrac{\pi}{2}+x\right)\text{vers}\left(\dfrac{\pi}{2}-x\right).$

$\dfrac{dy}{dx} = -\sin 2x.$

16. $y = \dfrac{e^x(a\sin x - \cos x)}{a^2+1}.$ $\quad \dfrac{dy}{dx} = e^x \sin x.$

17. $y = x^{\sin x}.$ $\quad \dfrac{dy}{dx} = y\left(\dfrac{\sin x}{x} + \cos x \log x\right).$

18. $y = \sin nx \sin^n x.$ $\quad \dfrac{dy}{dx} = n\sin^{n-1}x \sin(n+1)x.$

19. $y = \dfrac{\sin^m nx}{\cos^n mx}.$ $\quad \dfrac{dy}{dx} = \dfrac{mn\sin^{m-1}nx \cos(m-n)x}{\cos^{n+1}mx}.$

20. $y = x + \log\cos\left(x - \dfrac{\pi}{4}\right).$ $\quad \dfrac{dy}{dx} = \dfrac{2}{1+\tan x}.$

21. $y = \log\tan\left(\dfrac{x}{2} + \dfrac{\pi}{4}\right).$ $\quad \dfrac{dy}{dx} = \sec x.$

22. $y = \log\sqrt{\dfrac{1-\cos x}{1+\cos x}}.$ $\quad \dfrac{dy}{dx} = \csc x.$

23. $y = \log\sqrt{\dfrac{a\cos x - b\sin x}{a\cos x + b\sin x}}.$ $\quad \dfrac{dy}{dx} = \dfrac{-ab}{a^2\cos^2 x - b^2\sin^2 x}.$

24. $y = \dfrac{\tan x - \tan^3 x}{\sec^4 x}.$ $\quad \dfrac{dy}{dx} = \cos 4x.$

In each of the following pairs of equations derive by differentiation each of the two equations from the other:

25. $\sin 2x = 2\sin x \cos x,$
$\cos 2x = \cos^2 x - \sin^2 x.$

26. $\sin 2x = \dfrac{2\tan x}{1+\tan^2 x},$
$\cos 2x = \dfrac{1-\tan^2 x}{1+\tan^2 x}.$

27. $\sin 3x = 3\sin x - 4\sin^3 x,$
    $\cos 3x = 4\cos^3 x - 3\cos x.$

28. $\sin 4x = 4\sin x \cos^3 x - 4\cos x \sin^3 x,$
    $\cos 4x = 1 - 8\sin^2 x \cos^2 x.$

29. $\sin(m+n)x = \sin mx \cos nx + \cos mx \sin nx,$
    $\cos(m+n)x = \cos mx \cos nx - \sin mx \sin nx.$

30. $\sin x = x - \dfrac{x^3}{\underline{|3}} + \dfrac{x^5}{\underline{|5}} - \dfrac{x^7}{\underline{|7}} + \cdots$

    $\cos x = 1 - \dfrac{x^2}{\underline{|2}} + \dfrac{x^4}{\underline{|4}} - \dfrac{x^6}{\underline{|6}} + \cdots.$

31. $\sin x = \dfrac{e^{x\sqrt{-1}} - e^{-x\sqrt{-1}}}{2\sqrt{-1}},$

    $\cos x = \dfrac{e^{x\sqrt{-1}} + e^{-x\sqrt{-1}}}{2}.$

19. *Formulæ for Differentiation of Inverse Trigonometric Functions.*

$$\text{XX.} \quad \frac{d}{dx}\sin^{-1}u = \frac{\frac{du}{dx}}{\sqrt{1-u^2}}.$$

$$\text{XXI.} \quad \frac{d}{dx}\cos^{-1}u = -\frac{\frac{du}{dx}}{\sqrt{1-u^2}}.$$

$$\text{XXII.} \quad \frac{d}{dx}\tan^{-1}u = \frac{\frac{du}{dx}}{1+u^2}.$$

$$\text{XXIII.} \quad \frac{d}{dx}\cot^{-1}u = -\frac{\frac{du}{dx}}{1+u^2}.$$

$$\text{XXIV.} \quad \frac{d}{dx}\sec^{-1}u = \frac{\frac{du}{dx}}{u\sqrt{u^2-1}}.$$

XXV. $\quad \dfrac{d}{dx}\operatorname{cosec}^{-1}u = -\dfrac{\dfrac{du}{dx}}{u\sqrt{u^2-1}}.$

XXVI. $\quad \dfrac{d}{dx}\operatorname{vers}^{-1}u = \dfrac{\dfrac{du}{dx}}{\sqrt{2u-u^2}}.$

20. Derivation of Formulæ.

Proof of XX. Let $y = \sin^{-1}u$;

therefore $\quad \sin y = u.$

By XIII., $\quad \cos y \dfrac{dy}{dx} = \dfrac{du}{dx};$

therefore $\quad \dfrac{dy}{dx} = \dfrac{\dfrac{du}{dx}}{\cos y}.$

But $\quad \cos y = \sqrt{1-\sin^2 y} = \sqrt{1-u^2};$

therefore $\quad \dfrac{dy}{dx} = \dfrac{\dfrac{du}{dx}}{\sqrt{1-u^2}}.$

Proof of XXI. This may be derived like XX., or from the relation
$$\cos^{-1}u = \dfrac{\pi}{2} - \sin^{-1}u;$$

whence $\quad \dfrac{d}{dx}\cos^{-1}u = -\dfrac{d}{dx}\sin^{-1}u = -\dfrac{\dfrac{du}{dx}}{\sqrt{1-u^2}}.$

Proof of XXII. Let $y = \tan^{-1}u$;

therefore $\quad \tan y = u.$

By XV., $\quad \sec^2 y \dfrac{dy}{dx} = \dfrac{du}{dx};$

therefore $\quad \dfrac{dy}{dx} = \dfrac{\dfrac{du}{dx}}{\sec^2 y}.$

But $\sec^2 y = 1 + \tan^2 y = 1 + u^2$;

therefore $\dfrac{dy}{dx} = \dfrac{\dfrac{du}{dx}}{1 + u^2}.$

Proof of XXIII. This may be derived like XXII., or from the relation
$$\cot^{-1} u = \frac{\pi}{2} - \tan^{-1} u.$$

Proof of XXIV. Let $y = \sec^{-1} u$;

therefore $\sec y = u.$

By XVII., $\sec y \tan y \dfrac{dy}{dx} = \dfrac{du}{dx}$;

therefore $\dfrac{dy}{dx} = \dfrac{\dfrac{du}{dx}}{\sec y \tan y}.$

But $\sec y \tan y = \sec y \sqrt{\sec^2 y - 1} = u \sqrt{u^2 - 1}$;

therefore $\dfrac{dy}{dx} = \dfrac{\dfrac{du}{dx}}{u \sqrt{u^2 - 1}}.$

Proof of XXV. This may be derived like XXIV., or from the relation
$$\operatorname{cosec}^{-1} u = \frac{\pi}{2} - \sec^{-1} u.$$

Proof of XXVI. Let $y = \operatorname{vers}^{-1} u$;

therefore $u = \operatorname{vers} y = 1 - \cos y.$

By XIV., $\dfrac{du}{dx} = \sin y \dfrac{dy}{dx}$;

therefore $\dfrac{dy}{dx} = \dfrac{\dfrac{du}{dx}}{\sin y}.$

But $\sin y = \sqrt{1 - \cos^2 y} = \sqrt{1 - (1 - u)^2} = \sqrt{2u - u^2}$;

therefore $\dfrac{dy}{dx} = \dfrac{\dfrac{du}{dx}}{\sqrt{2u - u^2}}.$

DIFFERENTIATION.

EXAMPLES.

1. $y = \tan^{-1} mx.$ $\qquad \dfrac{dy}{dx} = \dfrac{m}{1 + m^2 x^2}.$

2. $y = \sin^{-1}(3x - 1).$ $\qquad \dfrac{dy}{dx} = \dfrac{3}{\sqrt{6x - 9x^2}}.$

3. $y = \operatorname{vers}^{-1} \dfrac{8x}{9}.$ $\qquad \dfrac{dy}{dx} = \dfrac{2}{\sqrt{9x - 4x^2}}.$

4. $y = \sin^{-1}(3x - 4x^3).$ $\qquad \dfrac{dy}{dx} = \dfrac{3}{\sqrt{1 - x^2}}.$

5. $y = \tan^{-1} \dfrac{2x}{1 - x^2}.$ $\qquad \dfrac{dy}{dx} = \dfrac{2}{1 + x^2}.$

6. $y = \tan^{-1} e^x.$ $\qquad \dfrac{dy}{dx} = \dfrac{1}{e^x + e^{-x}}.$

7. $y = \tan^{-1}(n \tan x).$ $\qquad \dfrac{dy}{dx} = \dfrac{n}{\cos^2 x + n^2 \sin^2 x}.$

8. $y = \operatorname{cosec}^{-1} \dfrac{3}{2x}.$ $\qquad \dfrac{dy}{dx} = \dfrac{2}{\sqrt{9 - 4x^2}}.$

9. $y = \operatorname{vers}^{-1} 2x^2.$ $\qquad \dfrac{dy}{dx} = \dfrac{2}{\sqrt{1 - x^2}}.$

10. $y = \operatorname{vers}^{-1} \dfrac{2x^2}{1 + x^2}.$ $\qquad \dfrac{dy}{dx} = \dfrac{2}{1 + x^2}.$

11. $y = \tan^{-1} \dfrac{e^x - e^{-x}}{2}.$ $\qquad \dfrac{dy}{dx} = \dfrac{2}{e^x + e^{-x}}.$

12. $y = \operatorname{cosec}^{-1} \dfrac{1}{2x^2 - 1}.$ $\qquad \dfrac{dy}{dx} = \dfrac{2}{\sqrt{1 - x^2}}.$

13. $y = \sec^{-1} \dfrac{x^2 + 1}{x^2 - 1}.$ $\qquad \dfrac{dy}{dx} = \dfrac{-2}{x^2 + 1}.$

14. $y = \sin^{-1}\dfrac{x+1}{\sqrt{2}}$. $\dfrac{dy}{dx} = \dfrac{1}{\sqrt{1-2x-x^2}}$.

15. $y = \tan^{-1}\dfrac{4\sin x}{3+5\cos x}$. $\dfrac{dy}{dx} = \dfrac{4}{5+3\cos x}$.

16. $y = \cos^{-1}\dfrac{3+5\cos x}{5+3\cos x}$. $\dfrac{dy}{dx} = \dfrac{4}{5+3\cos x}$.

17. $y = \sin^{-1}\dfrac{1-x^2}{1+x^2}$. $\dfrac{dy}{dx} = \dfrac{-2}{1+x^2}$.

18. $y = \operatorname{cosec}^{-1}\dfrac{1+x^2}{2x}$. $\dfrac{dy}{dx} = \dfrac{2}{1+x^2}$.

19. $y = \tan^{-1}\dfrac{x+a}{1-ax}$. $\dfrac{dy}{dx} = \dfrac{1}{1+x^2}$.

20. $y = \sin^{-1}\sqrt{\sin x}$. $\dfrac{dy}{dx} = \dfrac{1}{2}\sqrt{1+\operatorname{cosec} x}$.

21. $y = \tan^{-1}\sqrt{\dfrac{1-\cos x}{1+\cos x}}$. $\dfrac{dy}{dx} = \dfrac{1}{2}$.

22. $y = \tan^{-1}\dfrac{\sqrt{x}+\sqrt{a}}{1-\sqrt{ax}}$. $\dfrac{dy}{dx} = \dfrac{1}{2\sqrt{x}(1+x)}$.

23. $y = \cot^{-1}\dfrac{a}{x} + \log\sqrt{\dfrac{x-a}{x+a}}$. $\dfrac{dy}{dx} = \dfrac{2ax^2}{x^4-a^4}$.

24. $y = \tan^{-1}(x+\sqrt{1-x^2})$. $\dfrac{dy}{dx} = \dfrac{\sqrt{1-x^2}-x}{2\sqrt{1-x^2}(1+x\sqrt{1-x^2})}$.

25. $y = \cos^{-1}\dfrac{e^x-e^{-x}}{e^x+e^{-x}}$. $\dfrac{dy}{dx} = \dfrac{-2}{e^x+e^{-x}}$.

26. $y = \sec^{-1}\sqrt{\dfrac{2}{1+x}}$. $\dfrac{dy}{dx} = \dfrac{-1}{2\sqrt{1-x^2}}$.

27. $y = (x+a)\tan^{-1}\sqrt{\dfrac{x}{a}} - \sqrt{ax}$. $\dfrac{dy}{dx} = \tan^{-1}\sqrt{\dfrac{x}{a}}$.

DIFFERENTIATION.

28. $y = \cot^{-1}\dfrac{1+\sqrt{1+x^2}}{x}.$ $\qquad \dfrac{dy}{dx} = \dfrac{1}{2(1+x^2)}.$

29. $y = \sin^{-1}\dfrac{x\tan a}{\sqrt{a^2-x^2}}.$ $\qquad \dfrac{dy}{dx} = \dfrac{a^2\tan a}{a^2-x^2}\dfrac{1}{\sqrt{a^2-x^2\sec^2 a}}.$

30. $y = \cot^{-1}\sqrt{\dfrac{1-x}{2+x}}.$ $\qquad \dfrac{dy}{dx} = \dfrac{1}{2\sqrt{2-x-x^2}}.$

31. $y = \tan^{-1}\dfrac{3x-x^3}{1-3x^2}.$ $\qquad \dfrac{dy}{dx} = \dfrac{3}{1+x^2}.$

32. $y = \tan^{-1}\dfrac{2x}{1+3x^2} + \cot^{-1}\dfrac{x}{1+2x^2} + \tan^{-1}2x.$ $\dfrac{dy}{dx} = \dfrac{3}{1+9x^2}.$

33. $y = \tan^{-1}\dfrac{2x-b}{b\sqrt{3}} + \tan^{-1}\dfrac{2b-x}{x\sqrt{3}}.$ $\qquad \dfrac{dy}{dx} = 0.$

34. $y = \log\sqrt{\dfrac{2x^2-2x+1}{2x^2+2x+1}} + \tan^{-1}\dfrac{2x}{1-2x^2}.$ $\dfrac{dy}{dx} = \dfrac{8x^2}{4x^4+1}.$

21. *To express* $\dfrac{dy}{dx}$ *in terms of* $\dfrac{dx}{dy}$. If y is a function of x, then (Art. 2) x may be regarded as a function of y. From the former relation we have $\dfrac{dy}{dx}$, and from the latter, $\dfrac{dx}{dy}$. These differential coefficients are connected by a simple relation.

It is evident that $\qquad \dfrac{\Delta y}{\Delta x} = \dfrac{1}{\dfrac{\Delta x}{\Delta y}},$

however small the values of Δx and Δy. As these quantities approach zero, we have, for the limits of the members of this equation,

$$\dfrac{dy}{dx} = \dfrac{1}{\dfrac{dx}{dy}}. \qquad \cdots \cdots \cdots \quad (1)$$

That is, the relation between $\dfrac{dy}{dx}$ and $\dfrac{dx}{dy}$ is the same as if they were ordinary fractions.

For example, suppose

$$x = \frac{a}{y+1}. \quad \ldots \ldots \ldots \quad (2)$$

Differentiating with respect to y, we have

$$\frac{dx}{dy} = -\frac{a}{(y+1)^2}.$$

By (1), $\quad \dfrac{dy}{dx} = -\dfrac{(y+1)^2}{a} = -\dfrac{a}{x^2}, \quad$ by (2).

This is the same result that we get by solving (2) with reference to y, giving

$$y = \frac{a}{x} - 1,$$

and differentiating this with reference to x.

22. *To express $\dfrac{dy}{dx}$ in terms of $\dfrac{dy}{dz}$ and $\dfrac{dz}{dx}$.* If y is a given function of z, and z a given function of x, it follows that y is a function of x. This relation may be obtained by eliminating z between the two given equations, but $\dfrac{dy}{dx}$ can be found without such elimination.

By differentiating the two given equations, we find $\dfrac{dy}{dz}$ and $\dfrac{dz}{dx}$, and from these differential coefficients, $\dfrac{dy}{dx}$ may be obtained by a relation which may be derived as follows:

It is evident that $\quad \dfrac{\Delta y}{\Delta x} = \dfrac{\Delta y}{\Delta z} \dfrac{\Delta z}{\Delta x},$

however small Δx, Δy, and Δz. As these quantities approach zero, we have for the limits of the members of this equation,

$$\frac{dy}{dx} = \frac{dy}{dz}\frac{dz}{dx}. \quad \ldots \ldots \ldots \quad (1)$$

That is, the relation is the same as if the differential coefficients were ordinary fractions.

For example, suppose

$$y = z^5, \quad z = a^2 - x^2. \qquad \qquad (2)$$

Differentiating these equations, the first with reference to z, and the second with reference to x, we have

$$\frac{dy}{dz} = 5z^4, \qquad \frac{dz}{dx} = -2x.$$

By (1), $\quad \dfrac{dy}{dx} = 5z^4(-2x) = -10x(a^2-x^2)^4, \quad$ by (2).

The same result might have been obtained by eliminating z between (2), giving

$$y = (a^2 - x^2)^5,$$

and differentiating this with reference to x.

EXAMPLES.

In the following seven examples find by differentiation $\dfrac{dx}{dy}$, and then $\dfrac{dy}{dx}$ by (1) Art. 21.

1. $x = \dfrac{2y}{y-1}.$ $\quad \dfrac{dy}{dx} = -\dfrac{(y-1)^2}{2} = -\dfrac{2}{(x-2)^2}.$

2. $x = \sqrt{y^2+1} - y.$ $\quad \dfrac{dy}{dx} = \dfrac{\sqrt{y^2+1}}{y - \sqrt{y^2+1}} = -\dfrac{x^2+1}{2x^2}.$

3. $x = \sqrt{1+\sin y}.$ $\quad \dfrac{dy}{dx} = \dfrac{2\sqrt{1+\sin y}}{\cos y} = \dfrac{2}{\sqrt{2-x^2}}.$

4. $x = \tan^{-1}(y + \sqrt{y^2-1}).$ $\dfrac{dy}{dx} = 2y\sqrt{y^2-1} = \dfrac{1}{2}(\tan^2 x - \cot^2 x).$

5. $x = \dfrac{y}{1+\log y}.$ $\quad \dfrac{dy}{dx} = \dfrac{(1+\log y)^2}{\log y} = \dfrac{y^2}{xy - x^2}.$

6. $x = \log \dfrac{e^y + \sqrt{e^{2y}-4}}{2}.$ $\quad \dfrac{dy}{dx} = \dfrac{\sqrt{e^{2y}-4}}{e^y} = \dfrac{e^{2x}-1}{e^{2x}+1}.$

7. $x = 2\log\dfrac{\sqrt{e^y+2}+\sqrt{e^y-2}}{2}$. $\dfrac{dy}{dx} = \dfrac{\sqrt{e^{2y}-4}}{e^y} = \dfrac{e^{2x}-1}{e^{2x}+1}$.

In the following examples find by differentiation $\dfrac{dy}{dz}$ and $\dfrac{dz}{dx}$, and then $\dfrac{dy}{dx}$ by (1) Art. 22.

8. $y = \dfrac{2z}{3z-2}$, $z = \dfrac{x}{2x-1}$. $\dfrac{dy}{dx} = \dfrac{4}{(x-2)^2}$.

9. $y = e^z + e^{2z}$, $z = \log(x-x^2)$. $\dfrac{dy}{dx} = 4x^3 - 6x^2 + 1$.

10. $y = \log(z^{\frac{3}{2}} - z)$, $z = e^{3x}$. $\dfrac{dy}{dx} = \dfrac{5e^{2x}-3}{e^{2x}-1}$.

11. $y = \log\dfrac{1+z^2}{z}$, $z = e^x$. $\dfrac{dy}{dx} = \dfrac{e^x - e^{-x}}{e^x + e^{-x}}$.

12. $y = \tan 2z$, $z = \tan^{-1}(2x-1)$. $\dfrac{dy}{dx} = \dfrac{2x^2 - 2x + 1}{2(x-x^2)^2}$.

13. $y = \dfrac{1}{6}\log\dfrac{(z+1)^2}{z^2 - z + 1} - \dfrac{1}{\sqrt{3}}\tan^{-1}\dfrac{2z-1}{\sqrt{3}}$, $z = \dfrac{\sqrt[3]{1+3x+3x^2}}{x}$.

$\dfrac{dy}{dx} = \dfrac{1}{xz(1+x)}$.

CHAPTER IV.

SUCCESSIVE DIFFERENTIATION.

23. *Definition*. A single differentiation performed on $y = f(x)$ gives the differential coefficient, $\dfrac{dy}{dx}$. This result being generally also a function of x, may be again differentiated, and we thus obtain what is called the *second differential coefficient*; the result of three successive differentiations is the *third differential coefficient*; and so on.

For example, if
$$y = x^4,$$
$$\frac{dy}{dx} = 4x^3,$$
$$\frac{d}{dx}\frac{dy}{dx} = 12x^2,$$
$$\frac{d}{dx}\frac{d}{dx}\frac{dy}{dx} = 24x.$$

24. *Notation*. The second differential coefficient of y with respect to x, is denoted by $\dfrac{d^2y}{dx^2}$.

That is, $\quad \dfrac{d^2y}{dx^2} = \dfrac{d}{dx}\dfrac{dy}{dx}.$

Similarly, $\quad \dfrac{d^3y}{dx^3} = \dfrac{d}{dx}\dfrac{d}{dx}\dfrac{dy}{dx} \; = \dfrac{d}{dx}\dfrac{d^2y}{dx^2}.$

$\quad\quad\quad\quad\;\; \dfrac{d^4y}{dx^4} = \dfrac{d}{dx}\dfrac{d}{dx}\dfrac{d}{dx}\dfrac{dy}{dx} = \dfrac{d}{dx}\dfrac{d^3y}{dx^3}.$

... ...

$\quad\quad\quad\quad\;\; \dfrac{d^n y}{dx^n} = \dfrac{d}{dx}\dfrac{d^{n-1}y}{dx^{n-1}}.$

Thus, if
$$y = x^4,$$
$$\frac{dy}{dx} = 4x^3,$$
$$\frac{d^2y}{dx^2} = 12x^2,$$
$$\frac{d^3y}{dx^3} = 24x.$$

The successive differential coefficients are sometimes called the *first, second, third, ··· derivatives*.

If the original function of x is denoted by $f(x)$, its successive differential coefficients are often denoted by

$$f'(x), \quad f''(x), \quad f'''(x), \quad \cdots \quad f^n(x).$$

25. *The nth Differential Coefficient.* It is possible to express the nth differential coefficient of some functions.

For example,

(a). From $y = e^x$, we have
$$\frac{dy}{dx} = e^x, \quad \frac{d^2y}{dx^2} = e^x, \quad \cdots \quad \frac{d^ny}{dx^n} = e^x.$$

(b). From $y = e^{ax}$, we have
$$\frac{dy}{dx} = ae^{ax}, \quad \frac{d^2y}{dx^2} = a^2 e^{ax}, \quad \cdots \quad \frac{d^ny}{dx^n} = a^n e^{ax}.$$

(c). From $y = \log x$, we have
$$\frac{dy}{dx} = x^{-1}, \quad \frac{d^2y}{dx^2} = (-1)x^{-2}, \quad \frac{d^3y}{dx^3} = (-1)(-2)x^{-3} = (-1)^2 \lfloor 2 \, x^{-3},$$
$$\frac{d^4y}{dx^4} = (-1)^3 \lfloor 3 \, x^{-4}, \quad \cdots \quad \frac{d^ny}{dx^n} = \frac{(-1)^{n-1} \lfloor n-1}{x^n}.$$

(d). From $y = \sin ax$, we have

$$\frac{dy}{dx} = a \cos ax = a \sin\left(ax + \frac{\pi}{2}\right),$$

$$\frac{d^2y}{dx^2} = a^2 \cos\left(ax + \frac{\pi}{2}\right) = a^2 \sin\left(ax + \frac{2\pi}{2}\right),$$

$$\frac{d^3y}{dx^3} = a^3 \cos\left(ax + \frac{2\pi}{2}\right) = a^3 \sin\left(ax + \frac{3\pi}{2}\right),$$

... ...

$$\frac{d^n y}{dx^n} = a^n \sin\left(ax + \frac{n\pi}{2}\right).$$

EXAMPLES.

1. $y = x^4 - 4x^3 + 6x^2 - 4x + 1.$ $\dfrac{d^2y}{dx^2} = 12(x^2 - 2x + 1).$

2. $y = x^5.$ $\dfrac{d^5y}{dx^5} = \lfloor 5.$

3. $y = (x-3)e^{2x} + 4xe^x + x.$ $\dfrac{d^2y}{dx^2} = 4e^x[(x-2)e^x + x + 2].$

4. $y = \dfrac{a}{x^m}.$ $\dfrac{d^2y}{dx^2} = \dfrac{m(m+1)a}{x^{m+2}}.$

5. $y = x \log x.$ $\dfrac{d^2y}{dx^2} = \dfrac{1}{x}.$

6. $y = x^3 \log x.$ $\dfrac{d^4y}{dx^4} = \dfrac{6}{x}.$

7. $y = \log(e^x + e^{-x}).$ $\dfrac{d^3y}{dx^3} = -\dfrac{8(e^x - e^{-x})}{(e^x + e^{-x})^3}.$

8. $y = (x^2 - 6x + 12)e^x.$ $\dfrac{d^3y}{dx^3} = x^2 e^x.$

9. $y = \dfrac{x^3}{6}\left(\log x - \dfrac{5}{6}\right).$ $\dfrac{d^2y}{dx^2} = x \log x.$

10. $y = \log \sin x.$ $\dfrac{d^3y}{dx^3} = \dfrac{2 \cos x}{\sin^3 x}.$

11. $y = (x^2 + a^2) \tan^{-1} \dfrac{x}{a}.$ $\dfrac{d^3y}{dx^3} = \dfrac{4 a^3}{(a^2 + x^2)^2}.$

12. $y = e^{-x} \cos x.$ $\dfrac{d^4y}{dx^4} = -4 e^{-x} \cos x.$

13. $y = \tan x.$ $\dfrac{d^3y}{dx^3} = 6 \sec^4 x - 4 \sec^2 x.$

14. $y = \dfrac{5x+1}{x^2-1}.$ $\dfrac{d^6y}{dx^6} = \lfloor 6 \left[\dfrac{3}{(x-1)^7} + \dfrac{2}{(x+1)^7}\right].$

Decompose the fraction before differentiating.

15. $y = \sqrt{\sec 2x}.$ $\dfrac{d^2y}{dx^2} = 3 y^5 - y.$

16. $y = (e^x + e^{-x})^n.$ $\dfrac{d^2y}{dx^2} = n^2 y - 4n(n-1) y^{\frac{n-2}{n}}.$

17. $y = \dfrac{7 \cos x}{9} - \dfrac{\cos^3 x}{27}.$ $\dfrac{d^3y}{dx^3} = \sin^3 x.$

18. $y = \tan^2 x + 8 \log \cos x + 3 x^2.$ $\dfrac{d^2y}{dx^2} = 6 \tan^4 x.$

19. $y = (x^2 - 3x + 3) e^{2x}.$ $\dfrac{d^3y}{dx^3} = 8 x^2 e^{2x}.$

20. $y = x^3 \left[3 (\log x)^2 - 11 \log x + \dfrac{85}{6}\right].$ $\dfrac{d^3y}{dx^3} = 18 (\log x)^2.$

21. $y = e^{ax} \sin bx.$ $\dfrac{d^2y}{dx^2} - 2 a \dfrac{dy}{dx} + (a^2 + b^2) y = 0.$

22. $y = \sin(m \sin^{-1} x).$ $(1 - x^2) \dfrac{d^2y}{dx^2} - x \dfrac{dy}{dx} + m^2 y = 0.$

SUCCESSIVE DIFFERENTIATION.

23. $y = a\cos(\log x) + b\sin(\log x).\quad x^2\dfrac{d^2y}{dx^2} + x\dfrac{dy}{dx} + y = 0.$

24. $y = \dfrac{1}{x+2}.\qquad\qquad \dfrac{d^n y}{dx^n} = \dfrac{(-1)^n \lfloor n}{(x+2)^{n+1}}.$

25. $y = \dfrac{1}{3x+4}.\qquad\qquad \dfrac{d^n y}{dx^n} = \dfrac{(-1)^n 3^n \lfloor n}{(3x+4)^{n+1}}.$

26. $y = a^x.\qquad\qquad\qquad \dfrac{d^n y}{dx^n} = (\log a)^n a^x.$

27. $y = \cos ax.\qquad\qquad \dfrac{d^n y}{dx^n} = a^n \cos\left(ax + \dfrac{n\pi}{2}\right).$

28. $y = \dfrac{1-x}{1+x}.\qquad\qquad \dfrac{d^n y}{dx^n} = \dfrac{2(-1)^n \lfloor n}{(1+x)^{n+1}}.$

Reduce the fraction to a mixed quantity, $-1 + \dfrac{2}{1+x}$, before differentiating.

29. $y = \dfrac{3x+2}{x^2-4}.\quad \dfrac{d^n y}{dx^n} = (-1)^n \lfloor n \left[\dfrac{1}{(x+2)^{n+1}} + \dfrac{2}{(x-2)^{n+1}}\right].$

26. Leibnitz's Theorem. This is a formula for the nth differential coefficient of the product of two variables in terms of the successive differential coefficients of those variables.

A special case of Leibnitz's Theorem, when $n = 1$, is formula IV.,

$$\dfrac{d}{dx}(uv) = \dfrac{du}{dx}v + u\dfrac{dv}{dx}. \quad\cdots\quad (1)$$

For convenience let us use the following abridged notation:

$$v_1 = \dfrac{dv}{dx},\quad v_2 = \dfrac{d^2v}{dx^2},\quad \cdots\quad v_n = \dfrac{d^n v}{dx^n}.$$

$$u_1 = \dfrac{du}{dx},\quad u_2 = \dfrac{d^2u}{dx^2},\quad \cdots\quad u_n = \dfrac{d^n u}{dx^n}.$$

Then (1) becomes

$$\dfrac{d}{dx}(uv) = u_1 v + u v_1. \quad\cdots\quad (2)$$

Differentiating (2),

$$\frac{d^2}{dx^2}(uv) = u_2 v + u_1 v_1 + u_1 v_1 + u v_2 = u_2 v + 2 u_1 v_1 + u v_2.$$

$$\frac{d^3}{dx^3}(uv) = u_3 v + u_2 v_1 + 2 u_2 v_1 + 2 u_1 v_2 + u_1 v_2 + u v_3$$

$$= u_3 v + 3 u_2 v_1 + 3 u_1 v_2 + u v_3.$$

We shall find this law of the terms to apply, however far we continue the differentiation, the coefficients being those of the Binomial Theorem.

In general

$$\frac{d^n}{dx^n}(uv) = u_n v + n u_{n-1} v_1 + \frac{n(n-1)}{\underline{|2}} u_{n-2} v_2 + \cdots$$

$$+ n u_1 v_{n-1} + u v_n. \quad \cdots \cdots \cdots (3)$$

This may be proved by induction, by showing that if true for $\frac{d^n}{dx^n}(uv)$, it is also true for $\frac{d^{n+1}}{dx^{n+1}}(uv)$. This exercise is left for the student.

In the ordinary notation (3) becomes

$$\frac{d^n}{dx^n}(uv) = \frac{d^n u}{dx^n} v + n \frac{d^{n-1} u}{dx^{n-1}} \frac{dv}{dx} + \frac{n(n-1)}{\underline{|2}} \frac{d^{n-2} u}{dx^{n-2}} \frac{d^2 v}{dx^2} + \cdots$$

$$+ n \frac{du}{dx} \frac{d^{n-1} v}{dx^{n-1}} + u \frac{d^n v}{dx^n}. \quad \cdots \cdots (4)$$

For example, let us find by Leibnitz's Theorem $\frac{d^n}{dx^n}(e^{ax} x)$.

Here $u = e^{ax}, \quad u_1 = a e^{ax}, \quad \cdots \quad u_n = a^n e^{ax}.$

$v = x, \quad v_1 = 1, \quad v_2 = 0, \quad v_3 = 0, \quad \cdots.$

Substituting in (3), we have

$$\frac{d^n}{dx^n}(x e^{ax}) = a^n e^{ax} x + n a^{n-1} e^{ax} = a^{n-1} e^{ax}(ax + n).$$

EXAMPLES.

Find by Leibnitz's Theorem the following differential co-efficients:

1. $y = x^3 \tan x$. $\quad \dfrac{d^3y}{dx^3} = 2x^3 \sec^2 x (3\tan^2 x + 1) + 18 x^2 \sec^2 x \tan x$
$\quad + 18 x \sec^2 x + 6 \tan x$.

2. $y = e^x \log x$. $\quad \dfrac{d^4y}{dx^4} = e^x \left(\log x + \dfrac{4}{x} - \dfrac{6}{x^2} + \dfrac{8}{x^3} - \dfrac{6}{x^4} \right)$.

3. $y = x^2 a^x$. $\quad \dfrac{d^n y}{dx^n} = a^x (\log a)^{n-2} [(x \log a + n)^2 - n]$.

4. $y = \dfrac{x^2 + 1}{(x+1)^3}$. $\quad \dfrac{d^n y}{dx^n} = (-1)^n \lfloor n \dfrac{(x-n)^2 + n + 1}{(x+1)^{n+3}}$.

CHAPTER V.

DIFFERENTIALS.

27. The differential coefficient $\frac{dy}{dx}$ has been defined, not as a fraction having a numerator and denominator, but as a single symbol representing the limiting value of $\frac{\Delta y}{\Delta x}$, as Δx and Δy approach zero. But there are some advantages in regarding the differential coefficient as an actual fraction, dx and dy being infinitely small increments of x and y, and called *differentials* of x and y. That is, dx is an *infinitely small* Δx, and dy an *infinitely small* Δy.

For instance, if we differentiate $y = x^2$, we obtain

$$\frac{dy}{dx} = 2x.$$

Using differentials, this result might be written

$$dy = 2x\,dx.$$

These are two forms of expressing the same relation. According to the first, —

The limit of the ratio of the increment of y to that of x, as these increments approach zero, is $2x$.

According to the second, —

An infinitely small increment of y is $2x$ times the corresponding infinitely small increment of x.

We have the same two forms of expressing other relations in mathematics.

For instance, we may say, —

"The limit of the ratio, $\frac{arc}{chord}$, as these quantities approach zero, is unity."

Or, —

"An infinitely small arc is equal to its chord."

The equation $dy = 2x\,dx$ may thus be used as a convenient substitute for
$$\frac{dy}{dx} = 2x.$$

We see also why $\frac{dy}{dx}$ or $2x$ is called the *differential coefficient*, for it is the *coefficient* of dx in the equation $dy = 2x\,dx$.

28. The formulæ for differentiation may be expressed in the form of differentials by omitting the dx in each member. Thus, IV. becomes
$$d(uv) = v\,du + u\,dv;$$

and XXII., $\qquad d\tan^{-1}u = \dfrac{du}{1 + u^2};$

and the others may be similarly expressed.

Differentiation by the new formulæ is substantially the same as by the old, differing only in using the symbol d instead of $\dfrac{d}{dx}$.

For example, take Ex. 5, p. 17.

$$dy = d\left(\frac{x+3}{x^2+3}\right) = \frac{(x^2+3)d(x+3) - (x+3)d(x^2+3)}{(x^2+3)^2}$$

$$= \frac{(x^2+3)\,dx - (x+3)\,2x\,dx}{(x^2+3)^2}$$

$$= \frac{(x^2+3-2x^2-6x)\,dx}{(x^2+3)^2} = \frac{(3-6x-x^2)\,dx}{(x^2+3)^2}.$$

Dividing by dx gives
$$\frac{dy}{dx} = \frac{3-6x-x^2}{(x^2+3)^2}.$$

29. Successive Differentials. Successive differential coefficients, $\dfrac{d^2y}{dx^2}$, $\dfrac{d^3y}{dx^3}$, $\cdots$, which have been defined as single symbols, may also be interpreted as fractions, the numerators, d^2y, d^3y, $\cdots$, denoting $d(dy)$, $d[d(dy)]$, $\cdots$, and called the second, third, $\cdots$, differentials of y, while the denominators are $(dx)^2$, $(dx)^3$, $\cdots$.

This will be better understood from an example.

Let
$$y = x^4,$$
then
$$dy = 4x^3 dx.$$

As $4x^3 dx$ is a variable, dy is a variable, and may be again differentiated. Now, x being the independent variable, its increment dx may be supposed the same infinitely small quantity for all values of x; that is, we may regard dx as constant in the preceding equation. Thus we obtain

$$d(dy) = 12 x^2 dx \cdot dx = 12 x^2 (dx)^2.$$

Denoting $d(dy)$ by d^2y,

$$d^2y = 12 x^2 (dx)^2.$$

Differentiating again, and still regarding dx as constant,

$$d(d^2y) = 24 x\, dx\, (dx)^2 = 24 x (dx)^3,$$
or
$$d^3y = 24 x (dx)^3.$$

From these equations, by dividing by the power of dx in the second members, we find

$$\frac{d^2y}{(dx)^2} = 12 x^2,$$

$$\frac{d^3y}{(dx)^3} = 24 x.$$

The independent variable x, whose differential is supposed constant, is sometimes called the *equicrescent* variable.

DIFFERENTIALS.

EXAMPLES.

Differentiate the following, using differentials in the process:

1. $y = \dfrac{x^2+2}{x+1}.$ $\qquad dy = \dfrac{x^2+2x-2}{(x+1)^2}dx.$

2. $y = \sqrt[n]{a^2+x^2}.$ $\qquad dy = \dfrac{2x}{n}(a^2+x^2)^{\frac{1-n}{n}}dx.$

3. $y = (e^x + e^{-x})^2.$ $\qquad dy = 2(e^{2x} - e^{-2x})dx.$

4. $y = e^x \log x.$ $\qquad dy = e^x\left(\log x + \dfrac{1}{x}\right)dx.$

5. $y = x - \dfrac{e^x - e^{-x}}{e^x + e^{-x}}.$ $\qquad dy = \left(\dfrac{e^x - e^{-x}}{e^x + e^{-x}}\right)^2 dx.$

6. $y = \sin^m x \cos^n x.$ $dy = \sin^{m-1}x \cos^{n-1}x(m\cos^2 x - n\sin^2 x)dx.$

7. $y = \dfrac{1}{3}\tan^3 x + \tan x.$ $\qquad dy = \sec^4 x\, dx.$

8. $y = \tan^{-1}\log x.$ $\qquad dy = \dfrac{dx}{x[1+(\log x)^2]}.$

CHAPTER VI.

IMPLICIT FUNCTIONS. (See also Art. 67.)

30. Hitherto, in finding $\frac{dy}{dx}, \frac{d^2y}{dx^2}, \frac{d^3y}{dx^3}, \ldots$, y has been an explicit function of x. When the relation between y and x is given by an equation containing these quantities but not solved with reference to y, y is said to be an *implicit* function of x.

If the equation can be solved with reference to y, we may find its differential coefficients by the methods already given. But this solution is not necessary for the differentiation, for by the use of the formulæ of differentiation we may derive $\frac{dy}{dx}, \frac{d^2y}{dx^2}, \frac{d^3y}{dx^3}, \ldots$, directly from the given equation.

31. For example, suppose the relation between y and x to be given by the equation

$$a^2y^2 + b^2x^2 = a^2b^2.$$

Differentiating with respect to x,

$$\frac{d}{dx}(a^2y^2 + b^2x^2) = 0,$$

$$2a^2y\frac{dy}{dx} + 2b^2x = 0,$$

$$\frac{dy}{dx} = -\frac{b^2x}{a^2y}.$$

Having thus obtained the first differential coefficient, we may, by differentiating again, derive the second differential coefficient.

$$\frac{d^2y}{dx^2} = -\frac{d}{dx}\frac{b^2x}{a^2y} = -\frac{a^2yb^2 - b^2xa^2\frac{dy}{dx}}{a^4y^2} = -\frac{b^2\left(y - x\frac{dy}{dx}\right)}{a^2y^2}.$$

Substituting now for $\frac{dy}{dx}$ its value,

$$\frac{d^2y}{dx^2} = -\frac{b^2(a^2y^2+b^2x^2)}{a^4y^3} = -\frac{b^4}{a^2y^3}.$$

By differentiating again we may obtain

$$\frac{d^3y}{dx^3} = -\frac{3b^6x}{a^4y^5}.$$

The first differentiation may be conveniently performed by differentials instead of differential coefficients. Thus we should have from the equation

$$a^2y^2 + b^2x^2 = a^2b^2,$$

$$2a^2y\,dy + 2b^2x\,dx = 0,$$

giving $\quad\quad \dfrac{dy}{dx} = -\dfrac{b^2x}{a^2y},\quad$ as before.

In deriving $\dfrac{d^2y}{dx^2},\ \dfrac{d^3y}{dx^3},\ \ldots$, it is better to use differential coefficients rather than differentials.

EXAMPLES.

1. $y^2 = 4ax.\quad\quad \dfrac{dy}{dx} = \dfrac{2a}{y},\ \dfrac{d^2y}{dx^2} = -\dfrac{4a^2}{y^3}.$

2. $\sin(xy) = mx.\quad \dfrac{dy}{dx} = \dfrac{m - y\cos(xy)}{x\cos(xy)}.$

3. $x^y = y^x.\quad \dfrac{dy}{dx} = \dfrac{y^2 - xy\log y}{x^2 - xy\log x} = \dfrac{y^2(1-\log x)}{x^2(1-\log y)}.$

4. $y^2 - 2xy = a^2.$

$\dfrac{dy}{dx} = \dfrac{y}{y-x},\ \dfrac{d^2y}{dx^2} = \dfrac{a^2}{(y-x)^3},\ \dfrac{d^3y}{dx^3} = -\dfrac{3a^2x}{(y-x)^5},\ \dfrac{d^2x}{dy^2} = -\dfrac{a^2}{y^3}.$

5. $y = \sin(x+y)$.

$$\frac{dy}{dx} = \frac{\cos(x+y)}{1-\cos(x+y)}, \quad \frac{d^2y}{dx^2} = \frac{-y}{[1-\cos(x+y)]^3}.$$

6. $e^{x+y} = xy$. $\quad \dfrac{dy}{dx} = -\dfrac{y(x-1)}{x(y-1)}, \quad \dfrac{d^2y}{dx^2} = -\dfrac{y[(x-1)^2+(y-1)^2]}{x^2(y-1)^3}.$

7. $\sec x \cos y = m$. $\quad \dfrac{dy}{dx} = \dfrac{\tan x}{\tan y}, \quad \dfrac{d^2y}{dx^2} = \dfrac{\tan^2 y - \tan^2 x}{\tan^3 y}.$

8. $x^3 + y^3 - 3axy = 0$. $\quad \dfrac{dy}{dx} = -\dfrac{x^2-ay}{y^2-ax}, \quad \dfrac{d^2y}{dx^2} = -\dfrac{2a^3xy}{(y^2-ax)^3}.$

9. $x = a - b\cos\theta, \; y = a\theta + b\sin\theta,$
the variables being x, y, and θ.

$$\frac{dy}{dx} = \frac{a+b\cos\theta}{b\sin\theta}, \quad \frac{d^2y}{dx^2} = -\frac{b+a\cos\theta}{b^2\sin^3\theta}.$$

CHAPTER VII.

EXPANSION OF FUNCTIONS.

32. The student is probably already familiar with methods of expanding certain functions into series. Thus, by ordinary division,

$$\frac{1}{1+x} = 1 - x + x^2 - x^3 + \cdots;$$

by the Binomial Theorem,

$$(a+x)^n = a^n + na^{n-1}x + \frac{n(n-1)}{\underline{|2}} a^{n-2}x^2 + \cdots.$$

But these methods are limited in their application to certain forms of functions. We are now about to consider a method of expansion applicable to all functions, and including as special cases the expansions just referred to.

These methods are known as *Taylor's Theorem* and *Maclaurin's Theorem*. These two theorems are so connected that either may be regarded as involving the other. We shall first consider Maclaurin's Theorem as the simpler in expression and derivation.

33. *Maclaurin's Theorem.* This is a theorem by which any function of x may be expanded into a series of terms arranged according to the ascending integral powers of x. It may be expressed as follows:

$$f(x) = f(0) + f'(0)\frac{x}{\underline{|1}} + f''(0)\frac{x^2}{\underline{|2}} + f'''(0)\frac{x^3}{\underline{|3}} + \cdots$$

in which $f(x)$ is the given function to be expanded, and $f'(x)$, $f''(x)$, $f'''(x)$, $\cdots$, its successive differential coefficients.

That is,
$$f'(x) = \frac{d}{dx}f(x),$$
$$f''(x) = \frac{d}{dx}f'(x),$$
$$f'''(x) = \frac{d}{dx}f''(x),$$
... ...

$f(0)$, $f'(0)$, $f''(0)$, ..., as the notation implies, denote the values of $f(x), f'(x), f''(x), \ldots$, when $x = 0$.

34. Derivation of Maclaurin's Theorem. This may be derived by the method of Indeterminate Coefficients by assuming

$$f(x) = A + Bx + Cx^2 + Dx^3 + Ex^4 + \cdots \quad \cdot \quad \cdot \quad \cdot \quad (1)$$

where $A, B, C, \cdots$ are supposed to be *constant* coefficients.

Differentiating successively, and using the notation just defined, we have

$$f'(x) = B + 2Cx + 3Dx^2 + 4Ex^3 + \cdots \quad \cdot \quad \cdot \quad \cdot \quad (2)$$
$$f''(x) = 2C + 2\cdot 3 Dx + 3\cdot 4 Ex^2 + \cdots \quad \cdot \quad \cdot \quad \cdot \quad (3)$$
$$f'''(x) = 2\cdot 3 D + 2\cdot 3\cdot 4 Ex + \cdots \quad \cdot \quad \cdot \quad \cdot \quad \cdot \quad (4)$$
$$f^{\text{iv}}(x) = 2\cdot 3\cdot 4 E + \cdots \quad \cdot \quad \cdot \quad \cdot \quad \cdot \quad \cdot \quad \cdot \quad (5)$$
... ...

Now since equation (1), and consequently (2), (3), ... are supposed true for all values of x, they will be true when $x = 0$. Substituting zero for x in these equations, we have

from (1), $f(0) = A$, or $A = f(0)$,

" (2), $f'(0) = B$, or $B = f'(0)$,

" (3), $f''(0) = 2C$, or $C = \dfrac{f''(0)}{\lfloor 2}$,

EXPANSION OF FUNCTIONS. 57

from (4), $f'''(0) = 2\cdot 3 D$, or $D = \dfrac{f'''(0)}{\lfloor 3}$,

" (5), $f^{iv}(0) = 2\cdot 3\cdot 4 E$, or $E = \dfrac{f^{iv}(0)}{\lfloor 4}$.

... ...

Substituting these values of A, B, C, $\cdots$ in (1), we have

$$f(x) = f(0) + f'(0)\dfrac{x}{1} + f''(0)\dfrac{x^2}{\lfloor 2} + f'''(0)\dfrac{x^3}{\lfloor 3} + \cdots \quad \cdot \quad \cdot \quad (6)$$

35. As an example in the application of Maclaurin's Theorem, let it be required to expand $\log(1+x)$ into a series.

$$f(x) = \log(1+x), \qquad f(0) = \log 1 = 0.$$

$$f'(x) = \dfrac{1}{1+x} = (1+x)^{-1}, \quad f'(0) = 1.$$

$$f''(x) = -(1+x)^{-2}, \qquad f''(0) = -1.$$

$$f'''(x) = 2(1+x)^{-3}, \qquad f'''(0) = 2.$$

$$f^{iv}(x) = -\lfloor 3(1+x)^{-4}, \qquad f^{iv}(0) = -\lfloor 3.$$

$$f^{v}(x) = \lfloor 4(1+x)^{-5}, \qquad f^{v}(0) = \lfloor 4.$$

... ...

Substituting in (6) Art. 34, we have

$$\log(1+x) = 0 + 1\cdot x - 1\cdot\dfrac{x^2}{2} + \dfrac{2x^3}{\lfloor 3} - \dfrac{\lfloor 3\, x^4}{\lfloor 4} + \dfrac{\lfloor 4\, x^5}{\lfloor 5} - \cdots$$

or $\quad \log(1+x) = x - \dfrac{x^2}{2} + \dfrac{x^3}{3} - \dfrac{x^4}{4} + \dfrac{x^5}{5} - \cdots.$

36. If, in the application of Maclaurin's Theorem to a given function, any of the quantities $f(0)$, $f'(0)$, $f''(0)$, $\cdots$ are infinite, this function is not capable of being expanded in the proposed series. This is the case with $\log x$, $x^{\frac{1}{2}}$, $\cot x$.

EXAMPLES.

Derive the following by Maclaurin's Theorem:

1. $\sin x = x - \dfrac{x^3}{\underline{|3}} + \dfrac{x^5}{\underline{|5}} - \dfrac{x^7}{\underline{|7}} + \cdots .$

2. $\cos x = 1 - \dfrac{x^2}{\underline{|2}} + \dfrac{x^4}{\underline{|4}} - \dfrac{x^6}{\underline{|6}} + \cdots .$

3. $e^x = 1 + x + \dfrac{x^2}{\underline{|2}} + \dfrac{x^3}{\underline{|3}} + \dfrac{x^4}{\underline{|4}} + \cdots .$

4. $(a+x)^n = a^n + na^{n-1}x + \dfrac{n(n-1)}{\underline{|2}}a^{n-2}x^2$
$\qquad\qquad + \dfrac{n(n-1)(n-2)}{\underline{|3}}a^{n-3}x^3 + \cdots .$

5. $\log_a(1+x) = M\left(x - \dfrac{x^2}{2} + \dfrac{x^3}{3} - \dfrac{x^4}{4} + \cdots\right)$, where $M = \log_a e$.

6. $\tan x = x + \dfrac{x^3}{3} + \dfrac{2x^5}{15} + \cdots .$

7. $\tan^{-1} x = x - \dfrac{x^3}{3} + \dfrac{x^5}{5} - \dfrac{x^7}{7} + \cdots .$

Here $\qquad f(x) = \tan^{-1} x,$

$$f'(x) = \dfrac{1}{1+x^2} = 1 - x^2 + x^4 - x^6 + \cdots,$$

$$f''(x) = -2x + 4x^3 - 6x^5 + \cdots,$$

$\qquad \cdots \qquad \cdots \qquad \cdots \qquad \cdots$

8. $\sin^{-1} x = x + \dfrac{1}{2}\cdot\dfrac{x^3}{3} + \dfrac{1\cdot 3}{2\cdot 4}\cdot\dfrac{x^5}{5} + \dfrac{1\cdot 3\cdot 5}{2\cdot 4\cdot 6}\cdot\dfrac{x^7}{7} + \cdots .$

Here $f(x) = \sin^{-1} x,$

$$f'(x) = \frac{1}{\sqrt{1-x^2}} = (1-x^2)^{-\frac{1}{2}}.$$

Expanding by the Binomial Theorem,

$$f'(x) = 1 + \frac{1}{2}x^2 + \frac{1\cdot 3}{2\cdot 4}x^4 + \frac{1\cdot 3\cdot 5}{2\cdot 4\cdot 6}x^6 + \cdots$$

$$= 1 + ax^2 + bx^4 + cx^6 + \cdots,$$

where $\quad a = \dfrac{1}{2}, \quad b = \dfrac{1\cdot 3}{2\cdot 4}, \quad c = \dfrac{1\cdot 3\cdot 5}{2\cdot 4\cdot 6}, \quad \cdots,$

$$f''(x) = 2ax + 4bx^3 + 6cx^5 + \cdots,$$

$\cdots \quad \cdots \quad \cdots \quad \cdots \quad \cdots$

9. $e^x \sec x = 1 + x + x^2 + \dfrac{2x^3}{3} + \cdots.$

10. $\log_{10} \cos x = - M\left(\dfrac{x^2}{2} + \dfrac{x^4}{12} + \dfrac{x^6}{45} + \cdots\right),$ where $M = .4342945.$

11. $\log(1 + \sin x) = x - \dfrac{x^2}{2} + \dfrac{x^3}{6} - \dfrac{x^4}{12} + \dfrac{x^5}{24} \cdots.$

12. From Ex. 7 derive

$$\frac{\pi}{4} = 1 - \frac{1}{3} + \frac{1}{5} - \frac{1}{7} + \frac{1}{9} - \cdots.$$

Also, since $\tan^{-1} 1 = \tan^{-1}\dfrac{1}{2} + \tan^{-1}\dfrac{1}{3},$

$$\frac{\pi}{4} = \frac{1}{2} - \frac{1}{3}\left(\frac{1}{2}\right)^3 + \frac{1}{5}\left(\frac{1}{2}\right)^5 - \frac{1}{7}\left(\frac{1}{2}\right)^7 + \cdots$$

$$+ \frac{1}{3} - \frac{1}{3}\left(\frac{1}{3}\right)^3 + \frac{1}{5}\left(\frac{1}{3}\right)^5 - \frac{1}{7}\left(\frac{1}{3}\right)^7 + \cdots$$

$$= .4636476 \cdots + .3217506 \cdots = .785398 \cdots.$$

$\therefore \pi = 3.141592 \cdots.$

The computation includes 10 terms of the first series and 7 of the second.

13. From Ex. 3 show that

$$e^{x\sqrt{-1}} = 1 - \frac{x^2}{\lfloor 2} + \frac{x^4}{\lfloor 4} - \cdots + \sqrt{-1}\left(x - \frac{x^3}{\lfloor 3} + \frac{x^5}{\lfloor 5} - \cdots\right)$$

$$= \cos x + \sqrt{-1}\sin x, \text{ by Exs. 1, 2.}$$

Similarly, show that

$$e^{-x\sqrt{-1}} = \cos x - \sqrt{-1}\sin x.$$

From these two equations derive the *exponential values* of the sine and cosine,

$$\sin x = \frac{e^{x\sqrt{-1}} - e^{-x\sqrt{-1}}}{2\sqrt{-1}},$$

$$\cos x = \frac{e^{x\sqrt{-1}} + e^{-x\sqrt{-1}}}{2}.$$

37. *Taylor's Theorem.* This is a theorem for expanding any function of the sum of two quantities in a series arranged according to the powers of one of these quantities.

As the Binomial Theorem expands $(x+h)^n$ in a series arranged according to the powers of h, so Taylor's Theorem expands *any function* of $(x+h)$ in a similar series. It may be expressed as follows:

$$f(x+h) = f(x) + f'(x)h + f''(x)\frac{h^2}{\lfloor 2} + f'''(x)\frac{h^3}{\lfloor 3} + \cdots.$$

38. The proof of Taylor's Theorem depends upon the following principle:

If we differentiate $f(x+h)$ with reference to x, regarding h constant, the result is the same as if we differentiate it with reference to h, regarding x constant.

EXPANSION OF FUNCTIONS.

That is, $\dfrac{d}{dx}f(x+h) = \dfrac{d}{dh}f(x+h)$.

For, let $z = x + h$,

then by (1) Art. 22,

$$\dfrac{d}{dx}f(x+h) = \dfrac{d}{dx}f(z) = \dfrac{d}{dz}f(z)\dfrac{dz}{dx},$$

$$\dfrac{d}{dh}f(x+h) = \dfrac{d}{dh}f(z) = \dfrac{d}{dz}f(z)\dfrac{dz}{dh}.$$

But $\dfrac{dz}{dx} = 1$, and $\dfrac{dz}{dh} = 1$;

therefore $\dfrac{d}{dx}f(x+h) = \dfrac{d}{dh}f(x+h)$.

39. *Derivation of Taylor's Theorem.* With the aid of the preceding article we can now derive Taylor's Theorem by the method of Indeterminate Coefficients. Assume

$$f(x+h) = A + Bh + Ch^2 + Dh^3 + \cdots \quad \ldots \quad (1)$$

where A, B, C, $\cdots$ are supposed to be functions of x but not of h.

Differentiating (1), first with reference to x, then with reference to h,

$$\dfrac{d}{dx}f(x+h) = \dfrac{dA}{dx} + \dfrac{dB}{dx}h + \dfrac{dC}{dx}h^2 + \dfrac{dD}{dx}h^3 + \cdots,$$

$$\dfrac{d}{dh}f(x+h) = B + 2Ch + 3Dh^2 + \cdots.$$

By Art. 38, the first members of these two equations are equal to each other, therefore

$$\dfrac{dA}{dx} + \dfrac{dB}{dx}h + \dfrac{dC}{dx}h^2 + \cdots = B + 2Ch + 3Dh^2 + \cdots.$$

Equating the coefficients of like powers of h according to the principle of Indeterminate Coefficients, we have

$$\frac{dA}{dx} = B, \qquad B = \frac{dA}{dx}.$$

$$\frac{dB}{dx} = 2C, \qquad C = \frac{1}{2}\frac{d^2A}{dx^2}.$$

$$\frac{dC}{dx} = 3D, \qquad D = \frac{1}{\underline{|3}}\frac{d^3A}{dx^3}.$$

... ...

The coefficient A may be found from (1) by putting $h = 0$, as the equation must hold for this value among others.

Then $\qquad A = f(x).$

Hence $\qquad B = \dfrac{dA}{dx} = f'(x).$

$$C = \frac{1}{2}\frac{d^2A}{dx^2} = \frac{1}{2}f''(x).$$

$$D = \frac{1}{\underline{|3}}\frac{d^3A}{dx^3} = \frac{1}{\underline{|3}}f'''(x).$$

... ...

Substituting these expressions for $A, B, C, \cdots$ in (1), we have

$$f(x+h) = f(x) + f'(x)h + f''(x)\frac{h^2}{\underline{|2}} + f'''(x)\frac{h^3}{\underline{|3}} + \cdots . \quad (2)$$

40. Maclaurin's Theorem may be obtained from Taylor's Theorem by substituting $x = 0$. We then have

$$f(h) = f(0) + f'(0)h + f''(0)\frac{h^2}{\underline{|2}} + f'''(0)\frac{h^3}{\underline{|3}} + \cdots .$$

This is Maclaurin's Theorem expressed in terms of h instead of x.

EXPANSION OF FUNCTIONS. 63

41. As an example in the application of Taylor's Theorem, let it be required to expand $\sin(x+h)$ into a series.

$$f(x+h) = \sin(x+h),$$
$$f(x) = \sin x,$$
$$f'(x) = \cos x,$$
$$f''(x) = -\sin x,$$
$$f'''(x) = -\cos x,$$
$$f^{iv}(x) = \sin x.$$
$$\ldots \quad \ldots \quad \ldots$$

Substituting these expressions in (2) Art. 39, we find
$$\sin(x+h) = \sin x + h\cos x - \frac{h^2}{\lfloor 2}\sin x - \frac{h^3}{\lfloor 3}\cos x + \frac{h^4}{\lfloor 4}\sin x + \cdots.$$

EXAMPLES.

Derive the following by Taylor's Theorem:

1. $\log(x+h) = \log x + \dfrac{h}{x} - \dfrac{h^2}{2x^2} + \dfrac{h^3}{3x^3} - \dfrac{h^4}{4x^4} + \cdots.$

2. $(x+h)^n = x^n + nx^{n-1}h + \dfrac{n(n-1)}{\lfloor 2}x^{n-2}h^2$
$$+ \dfrac{n(n-1)(n-2)}{\lfloor 3}x^{n-3}h^3 + \cdots.$$

3. $\cos(x+h) = \cos x - h\sin x - \dfrac{h^2}{\lfloor 2}\cos x + \dfrac{h^3}{\lfloor 3}\sin x + \cdots.$

4. $\tan(x+h) = \tan x + h\sec^2 x + h^2\sec^2 x \tan x$
$$+ \dfrac{h^3}{3}\sec^2 x (1 + 3\tan^2 x) + \cdots.$$

5. $e^{x+h} = e^x\left(1 + h + \dfrac{h^2}{\lfloor 2} + \dfrac{h^3}{\lfloor 3} + \cdots\right).$

6. $\log\sin(x+h) = \log\sin x + h\cot x - \dfrac{h^2}{2}\csc^2 x + \dfrac{h^3}{3}\dfrac{\cos x}{\sin^3 x} + \cdots.$

7. $\log \sec(x+h) = \log \sec x + h \tan x + \dfrac{h^2}{2}\sec^2 x$

$\qquad + \dfrac{h^3}{3}\sec^2 x \tan x + \dfrac{h^4}{12}\sec^2 x(1+3\tan^2 x) + \cdots.$

42. The preceding proofs of Taylor's and Maclaurin's Theorems by the method of Indeterminate Coefficients are not altogether satisfactory, inasmuch as the possibility of development in the proposed form is assumed.

Any rigorous proof of Taylor's Theorem, independent of Indeterminate Coefficients, is comparatively difficult. We give the following as presenting the least difficulties to the student.

43. *Continuous Functions.* A function is said to be continuous between certain values of the independent variable, when it changes *gradually* while the variable passes from one value to the other. In other words, a *continuous* function is one that can be represented by a *continuous* curve.

44. If a given function $\phi(x)$ is zero when $x=a$ and when $x=b$, and is finite and continuous between those values, as well as its differential coefficient $\phi'(x)$; then $\phi'(x)$ must be zero for some value of x between a and b.

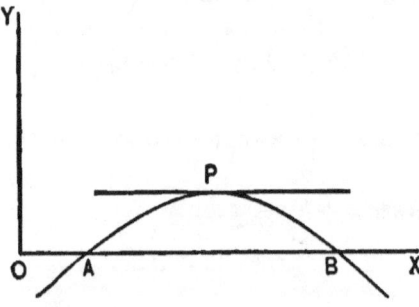

Let the function be represented by the curve $y=\phi(x)$. Let $OA=a$, $OB=b$. Then according to the hypothesis, $y=0$ when $x=a$, and when $x=b$.

Since the curve is continuous between A and B, there must be some point P between them, where the tangent is parallel to OX, and consequently $\phi'(x)=0$. (See Art. 94.) Hence the proposition is established.

EXPANSION OF FUNCTIONS. 65

With the aid of this proposition Taylor's Theorem can now be derived without the use of Indeterminate Coefficients.

45. *Proof of Taylor's Theorem.* Suppose $f(x)$ and its successive $n+1$ differential coefficients to be finite and continuous between $x = a$ and $x = a + h$. Let

$$\phi(x) = f(a+x) - f(a) - xf'(a) - \frac{x^2}{\lfloor 2}f''(a) \cdots - \frac{x^n}{\lfloor n}f^n(a) - \frac{x^{n+1}}{\lfloor n+1}R,$$

where

$$R = \frac{\lfloor n+1}{h^{n+1}}\left[f(a+h) - f(a) - hf'(a) - \frac{h^2}{\lfloor 2}f''(a) \cdots - \frac{h^n}{\lfloor n}f^n(a)\right].$$

It is to be noticed that R is independent of x.

It is evident that $\phi(x) = 0$ when $x = 0$ and when $x = h$. Hence by Art. 44, $\phi'(x) = 0$ for some value of x between 0 and h. Suppose h' this value. Then

$$\phi'(x) = f'(a+x) - f'(a) - xf''(a) - \frac{x^2}{\lfloor 2}f'''(a) \cdots - \frac{x^{n-1}}{\lfloor n-1}f^n(a)$$

$$-\frac{x^n}{\lfloor n}R = 0, \text{ when } x = h'.$$

But $\phi'(x) = 0$ when $x = 0$; hence $\phi''(x) = 0$ for some value of x between 0 and h'.

Continuing this process to $n + 1$ differentiations, we find

$$\phi^{n+1}(x) = f^{n+1}(a+x) - R = 0$$

for some value of x between 0 and h. Let this value of x be θh, where $\theta < 1$.

Then $\qquad f^{n+1}(a + \theta h) = R.$

Equating this value of R with that given above, we have

$$f(a+h) = f(a) + hf'(a) + \frac{h^2}{\lfloor 2}f''(a) \cdots + \frac{h^n}{\lfloor n}f^n(a)$$

$$+ \frac{h^{n+1}}{\lfloor n+1}f^{n+1}(a + \theta h).$$

We may now substitute x for a, since a may have *any* value, and we have

$$f(x+h) = f(x) + hf'(x) + \frac{h^2}{\lfloor 2} f''(x) \cdots + \frac{h^n}{\lfloor n} f^n(x)$$
$$+ \frac{h^{n+1}}{\lfloor n+1} f^{n+1}(x + \theta h).$$

46. *Remainder in Taylor's Theorem.* The last term

$$\frac{h^{n+1}}{\lfloor n+1} f^{n+1}(x + \theta h)$$

is called the remainder after $n+1$ terms. When the form of the function $f(x)$ is such that by taking n sufficiently large, this remainder can be made indefinitely small, then Taylor's Theorem gives a convergent series.

47. *Failure of Taylor's Theorem.* When $f(x)$ or any of its successive differential coefficients are infinite or discontinuous between x and $x+h$, the preceding demonstration no longer holds good, and for such a function Taylor's Theorem is said to fail.

48. *Remainder in Maclaurin's Theorem.* If we let $x = 0$ in the preceding equation, we have

$$f(h) = f(0) + hf'(0) + \frac{h^2}{\lfloor 2} f''(0) \cdots + \frac{h^n}{\lfloor n} f^n(0) + \frac{h^{n+1}}{\lfloor n+1} f^{n+1}(\theta h).$$

Or, substituting x for h,

$$f(x) = f(0) + xf'(0) + \frac{x^2}{\lfloor 2} f''(0) \cdots + \frac{x^n}{\lfloor n} f^n(0) + \frac{x^{n+1}}{\lfloor n+1} f^{n+1}(\theta x).$$

When the remainder, $\frac{x^{n+1}}{\lfloor n+1} f^{n+1}(\theta x)$, by taking n sufficiently large, can be made indefinitely small, the series is convergent.

49. Remainder in certain series. Let us apply the general expression for the remainder, $\dfrac{x^{n+1}}{\underline{|n+1}} f^{n+1}(\theta x)$, to the development of e^x. Here

$$R = \dfrac{x^{n+1}}{\underline{|n+1}} e^{\theta x}.$$

The fraction $\dfrac{x^{n+1}}{\underline{|n+1}}$ can be made as small as we please by taking n sufficiently large, whatever may be the value of x. Moreover, $e^{\theta x}$ is finite; hence R approaches zero.

Hence the series

$$e^x = 1 + x + \dfrac{x^2}{\underline{|2}} + \dfrac{x^3}{\underline{|3}} \cdots$$

is convergent for all values of x.

It is evident that $\dfrac{x^{n+1}}{\underline{|n+1}} f^{n+1}(\theta x)$ will have zero for its limit, whenever $f(x)$ is of such a form that all of its successive differential coefficients are finite. This is the case with $\sin x$ and $\cos x$. Hence these expansions

$$\sin x = x - \dfrac{x^3}{\underline{|3}} + \dfrac{x^5}{\underline{|5}} - \cdots,$$

$$\cos x = 1 - \dfrac{x^2}{\underline{|2}} + \dfrac{x^4}{\underline{|4}} - \cdots,$$

are convergent for all values of x.

If $f(x) = \log(1+x)$, then the remainder is

$$\dfrac{x^{n+1}}{\underline{|n+1}} \dfrac{(-1)^n \underline{|n}}{(1+\theta x)^{n+1}}.$$

This may be expressed as

$$R = \dfrac{(-1)^n}{n+1} \left(\dfrac{x}{1+\theta x} \right)^{n+1}.$$

If x is positive and equal to, or less than, unity, R has a limit of zero.

Hence the expansion

$$\log(1+x) = x - \frac{x^2}{2} + \frac{x^3}{3} - \frac{x^4}{4} + \cdots$$

is convergent for positive values of x, when $x=1$ or $x<1$, but divergent, when $x>1$.

CHAPTER VIII.

INDETERMINATE FORMS.

50. The value of a fraction is, in general, the value of the numerator divided by that of the denominator. When, however, the numerator and denominator being variable have, one or both, the value zero or infinity, the above definition is no longer applicable, and must be amended or enlarged.

The expression, *value of the fraction*, must be understood to mean, under these circumstances, *that value which the fraction approaches as its limit, when the numerator and denominator approach the assigned values.* We shall use it in this sense in the present chapter.

It is to be noticed that this new definition of the value of a fraction is not necessarily confined to the cases mentioned above, where the ordinary definition fails, but is of general application, since any value of a variable fraction may be regarded as a limiting value.

51. A fraction may take either of the three forms, $\frac{0}{a}, \frac{a}{0}, \frac{0}{0}$, (where a is a finite quantity), according as the numerator or denominator becomes zero, or both become zero.

In the first case, $\frac{0}{a} = 0$; that is, if the numerator approach zero, and the denominator a finite quantity, the fraction approaches zero as its limit.

In the second case, $\frac{a}{0} = \infty$; that is, when the numerator approaches a finite quantity, and the denominator zero, the fraction is increasing beyond any finite limit.

In the third case, $\frac{0}{0}$ is called *indeterminate*, for the reason that when both numerator and denominator approach zero, this alone is not sufficient to determine the limit of the fraction, which can only be found from the general form of the fraction.

For instance, consider the fraction $\dfrac{x^2-3x+2}{x^2-1}$.

When $x=2$, the fraction takes the form $\dfrac{0}{3}=0$.

When $x=-1$, the fraction takes the form $\dfrac{6}{0}=\infty$.

When $x=1$, the fraction takes the form $\dfrac{0}{0}$, which is indeterminate.

52. *To evaluate a fraction that takes the indeterminate form $\dfrac{0}{0}$.*

Frequently an algebraic transformation in the given fraction will determine the value. If the fraction in the preceding article be reduced to lower terms, its value, which was before indeterminate when $x=1$, will be found to be $-\dfrac{1}{2}$.

As another illustration, consider the fraction $\dfrac{x-2}{\sqrt{x-1}-1}$. When $x=2$, this takes the form $\dfrac{0}{0}$. But by rationalizing the denominator, we transform the fraction into

$$\frac{(x-2)(\sqrt{x-1}+1)}{x-2} = \sqrt{x-1}+1,$$

which becomes 2, when $x=2$.

53. The Differential Calculus furnishes the following method applicable to all cases.

Substitute for the numerator and denominator, respectively, their differential coefficients. The value of this new fraction for the assigned value of x will be the value required.

To prove this, suppose the fraction $\dfrac{\phi(x)}{\psi(x)} = \dfrac{0}{0}$, when $x=a$; that is, $\phi(a)=0$, and $\psi(a)=0$.

By Art. 50, the required value of the fraction is the limit of $\dfrac{\phi(a+h)}{\psi(a+h)}$, as h approaches zero.

INDETERMINATE FORMS. 71

By Taylor's Theorem,

$$\frac{\phi(x+h)}{\psi(x+h)} = \frac{\phi(x) + \phi'(x)h + \phi''(x)\frac{h^2}{\lfloor 2} + \phi'''(x)\frac{h^3}{\lfloor 3} + \cdots}{\psi(x) + \psi'(x)h + \psi''(x)\frac{h^2}{\lfloor 2} + \psi'''(x)\frac{h^3}{\lfloor 3} + \cdots}.$$

Substituting a for x, and remembering that $\phi(a) = 0$, $\psi(a) = 0$, we have

$$\frac{\phi(a+h)}{\psi(a+h)} = \frac{\phi'(a) + \phi''(a)\frac{h}{\lfloor 2} + \phi'''(a)\frac{h^2}{\lfloor 3} + \cdots}{\psi'(a) + \psi''(a)\frac{h}{\lfloor 2} + \psi'''(a)\frac{h^2}{\lfloor 3} + \cdots}; \quad \cdot \cdot \quad (1)$$

therefore, as h approaches zero,

the limit of $\dfrac{\phi(a+h)}{\psi(a+h)} = \dfrac{\phi'(a)}{\psi'(a)}$.

If $\phi'(a) = 0$, and $\psi'(a) = 0$, we have similarly from (1), as h approaches zero,

the limit of $\dfrac{\phi(a+h)}{\psi(a+h)} = \dfrac{\phi''(a)}{\psi''(a)}$;

that is, the process must be repeated, and as often as may be necessary to obtain a result which is not indeterminate.

For example, let us find the value of the fraction in Art. 51,

$$\frac{\phi(x)}{\psi(x)} = \frac{x^2 - 3x + 2}{x^2 - 1} = \frac{0}{0}, \quad \text{when } x = 1.$$

Hence $\dfrac{\phi'(x)}{\psi'(x)} = \dfrac{2x - 3}{2x} = -\dfrac{1}{2}$, when $x = 1$.

For another example, let us find the value of

$$\frac{\phi(x)}{\psi(x)} = \frac{e^x + e^{-x} - 2}{1 - \cos x} = \frac{0}{0}, \quad \text{when } x = 0.$$

$$\frac{\phi'(x)}{\psi'(x)} = \frac{e^x - e^{-x}}{\sin x} = \frac{0}{0}, \quad \text{when } x = 0.$$

$$\frac{\phi''(x)}{\psi''(x)} = \frac{e^x + e^{-x}}{\cos x} = 2, \quad \text{when } x = 0.$$

EXAMPLES.

Find the values of the following fractions:

1. $\dfrac{\log x}{x-1}$, when $x=1$. Ans. 1.

2. $\dfrac{x-2}{(x-1)^n - 1}$, when $x=2$. Ans. $\dfrac{1}{n}$.

3. $\dfrac{e^x - e^{-x}}{\sin x}$, $x=0$. Ans. 2.

4. $\dfrac{x \sin x}{x - 2\sin x}$, $x=0$. Ans. 0.

5. $\dfrac{\log(2x^2 - 1)}{\tan(x-1)}$, $x=1$. Ans. 4.

6. $\dfrac{\tan x - x}{x - \sin x}$, $x=0$. Ans. 2.

7. $\dfrac{\log \sin x}{(\pi - 2x)^2}$, $x=\dfrac{\pi}{2}$. Ans. $-\dfrac{1}{8}$.

8. $\dfrac{e^x - e^{-x} - 2x}{x - \sin x}$, $x=0$. Ans. 2.

9. $\dfrac{x^4 - 2x^3 + 2x - 1}{x^5 - 15x^2 + 24x - 10}$, $x=1$. Ans. $\dfrac{1}{10}$.

10. $\dfrac{2\tan x - \sin 2x}{\sin^3 x}$, $x=0$. Ans. 2.

11. $\dfrac{e^{5x} - 10 e^{2x+3} + 15 e^{x+4} - 6 e^5}{e^{4x} - 6 e^{2x+2} + 8 e^{x+3} - 3 e^4}$, $x=1$. Ans. $\dfrac{5e}{2}$.

12. $\dfrac{\sec^2 x - 2\tan x}{1 + \cos 4x}$, $x=\dfrac{\pi}{4}$. Ans. $\dfrac{1}{2}$.

13. $\dfrac{(e^x - e^2)^3}{(x-4)e^x + e^2 x}$, $x=2$. Ans. $6e^4$.

INDETERMINATE FORMS. 73

54. A fraction may take either of the forms, $\dfrac{\infty}{a}$, $\dfrac{a}{\infty}$, $\dfrac{\infty}{\infty}$.

By regarding the value of a fraction as a limit, it is evident that in the first two cases, $\dfrac{\infty}{a} = \infty$, and $\dfrac{a}{\infty} = 0$.

The form $\dfrac{\infty}{\infty}$ is indeterminate, for the reason that, if the numerator and denominator both increase beyond any finite limit, this alone is not sufficient to determine the limit of the fraction.

55. *To evaluate a fraction that takes the form* $\dfrac{\infty}{\infty}$.

Suppose $\dfrac{\phi(x)}{\psi(x)} = \dfrac{\infty}{\infty}$, when $x = a$;

that is, $\phi(a) = \infty$, and $\psi(a) = \infty$.

By taking the reciprocals of $\phi(x)$ and $\psi(x)$, we have

$$\frac{\phi(x)}{\psi(x)} = \frac{\dfrac{1}{\psi(x)}}{\dfrac{1}{\phi(x)}} = \frac{0}{0}, \text{ when } x = a.$$

Hence by Art. 53, the limiting value of $\dfrac{\phi(x)}{\psi(x)}$, when $x = a$, is the value of

$$\frac{\dfrac{d}{dx}\left(\dfrac{1}{\psi(x)}\right)}{\dfrac{d}{dx}\left(\dfrac{1}{\phi(x)}\right)} = \frac{-\dfrac{\psi'(x)}{[\psi(x)]^2}}{-\dfrac{\phi'(x)}{[\phi(x)]^2}} = \frac{\psi'(x)}{\phi'(x)}\left[\frac{\phi(x)}{\psi(x)}\right]^2, \text{ when } x = a.$$

That is, $\dfrac{\phi(a)}{\psi(a)} = \dfrac{\psi'(a)}{\phi'(a)}\left[\dfrac{\phi(a)}{\psi(a)}\right]^2;$ . . . . . . (1)

hence $1 = \dfrac{\psi'(a)\,\phi(a)}{\phi'(a)\,\psi(a)}$, or $\dfrac{\phi(a)}{\psi(a)} = \dfrac{\phi'(a)}{\psi'(a)}.$ . . (2)

In deriving (2), we have divided (1) by $\dfrac{\phi(a)}{\psi(a)}$. If, however, $\dfrac{\phi(a)}{\psi(a)} = 0$ or ∞, equation (2) does not logically follow from (1). Nevertheless, it may be shown that (2) is true in these cases also.

Suppose $\dfrac{\phi(a)}{\psi(a)} = 0$, and n a finite quantity,

then $\qquad \dfrac{\phi(a)}{\psi(a)} + n = \dfrac{\phi(a) + n\psi(a)}{\psi(a)} = n.$

To this last fraction, (2) evidently applies,

therefore $\qquad \dfrac{\phi(a) + n\psi(a)}{\psi(a)} = \dfrac{\phi'(a) + n\psi'(a)}{\psi'(a)};$

that is, $\dfrac{\phi(a)}{\psi(a)} + n = \dfrac{\phi'(a)}{\psi'(a)} + n,$ or $\dfrac{\phi(a)}{\psi(a)} = \dfrac{\phi'(a)}{\psi'(a)}.$

If $\qquad \dfrac{\phi(a)}{\psi(a)} = \infty,$ then $\dfrac{\psi(a)}{\phi(a)} = 0,$

and we have the preceding case.

Thus the form $\dfrac{\infty}{\infty}$ is evaluated in the same way as the form $\dfrac{0}{0}$.

For example, find the value of

$$\dfrac{\log x}{\cot x}, \text{ when } x = 0.$$

Here $\dfrac{\phi(x)}{\psi(x)} = \dfrac{\log x}{\cot x} = \dfrac{\infty}{\infty};$ $\qquad$ when $x = 0.$

$\dfrac{\phi'(x)}{\psi'(x)} = \dfrac{\dfrac{1}{x}}{-\operatorname{cosec}^2 x} = -\dfrac{\sin^2 x}{x} = \dfrac{0}{0},$ when $x = 0.$

$\dfrac{\phi''(x)}{\psi''(x)} = -\dfrac{2\sin x \cos x}{1} = \dfrac{0}{1} = 0,$ $\qquad$ when $x = 0.$

56. *To evaluate a function that takes the form $0 \cdot \infty$.*

The product $\phi(x) \cdot \psi(x)$ becomes indeterminate when one factor $= 0$, and the other $= \infty$.

INDETERMINATE FORMS. 75

By taking the reciprocal of one of the factors, the expression may be made to take the form $\frac{0}{0}$ or $\frac{\infty}{\infty}$.

For example, find the value of

$$(\pi - 2x)\tan x, \text{ when } x = \frac{\pi}{2}.$$

This takes the form $0 \cdot \infty$. But

$$(\pi - 2x)\tan x = \frac{\pi - 2x}{\cot x} = \frac{0}{0}, \text{ when } x = \frac{\pi}{2}.$$

The value is found by Art. 53 to be 2.

57. *To evaluate a function that takes the form* $\infty - \infty$.

Transform the expression into a fraction, which will assume either the form $\frac{0}{0}$ or $\frac{\infty}{\infty}$.

For example, find the value of

$$\frac{1}{\log x} - \frac{1}{x-1}, \text{ when } x = 1.$$

This takes the form $\infty - \infty$. But

$$\frac{1}{\log x} - \frac{1}{x-1} = \frac{x - 1 - \log x}{(x-1)\log x} = \frac{0}{0}, \text{ when } x = 1.$$

The value is found by Art. 53 to be $\frac{1}{2}$.

EXAMPLES.

Find the values of the following:

1. $\dfrac{\log\left(x - \dfrac{\pi}{2}\right)}{\tan x},$ when $x = \dfrac{\pi}{2}.$ Ans. 0.

2. $\sec 3x \cos 7x,$ $x = \dfrac{\pi}{2}.$ Ans. $\dfrac{7}{3}.$

3. $\sec x - \tan x,$ $x = \dfrac{\pi}{2}.$ Ans. 0.

76   DIFFERENTIAL CALCULUS.

4. $(a^{\frac{1}{x}} - 1)x,$ $x = \infty.$ Ans. $\log a.$

5. $\dfrac{\log \cot x}{\operatorname{cosec} x},$ $x = 0.$ Ans. $0.$

6. $\operatorname{cosec}^2 x - \dfrac{1}{x^2},$ $x = 0.$ Ans. $\dfrac{1}{3}.$

7. $\dfrac{e}{e^x - e} - \dfrac{1}{x-1},$ $x = 1.$ Ans. $-\dfrac{1}{2}.$

8. $(1 - \tan x)\sec 2x,$ $x = \dfrac{\pi}{4}.$ Ans. $1.$

9. $\dfrac{\sec \dfrac{\pi x}{2}}{\log(1-x)},$ $x = 1.$ Ans. $\infty.$

10. $(a^2 - x^2)\tan\dfrac{\pi x}{2a},$ $x = a.$ Ans. $\dfrac{4a^2}{\pi}.$

11. $\dfrac{\log \tan 2x}{\log \tan x},$ $x = \dfrac{\pi}{2}.$ Ans. $-1.$

12. $\dfrac{2}{\sin^2 x} - \dfrac{1}{1 - \cos x},$ $x = 0.$ Ans. $\dfrac{1}{2}.$

13. $2x\tan x - \pi \sec x,$ $x = \dfrac{\pi}{2}.$ Ans. $-2.$

14. $\dfrac{\tan\left[\dfrac{\pi}{4}(x+1)\right]}{\tan\dfrac{\pi x}{2}},$ $x = 1.$ Ans. $2.$

15. $\dfrac{\log\left(\sec\dfrac{\pi x}{2} + \tan\dfrac{\pi x}{2}\right)}{\log(x-1)},$ $x = 1.$ Ans. $-1.$

58. *To evaluate a function that takes either of the forms,* $0^0,\ \infty^0,\ 1^\infty.$

Take the logarithm of the given function, which will assume the form $0 \cdot \infty$, and can be evaluated by Art. 56. From this the value of the given function can be found.

INDETERMINATE FORMS.

For example, find the value of
$$(1+x)^{\frac{1}{x}} \text{ when } x = 0.$$
This takes the form 1^∞.

Let $$y = (1+x)^{\frac{1}{x}};$$
then $\log y = \dfrac{1}{x} \log(1+x) = \infty \cdot 0$, when $x = 0$.

The value of $\log y$ is found to be 1. Hence the value of y is e.

EXAMPLES.

Find the value of the following:

1. $(1+x^2)^{\frac{1}{x}}$, when $x = 0$. Ans. 1.

2. $(e^x + 1)^{\frac{1}{x}}$, $x = \infty$. Ans. e.

3. $(\cos 2x)^{\frac{1}{x^2}}$, $x = 0$. Ans. $\dfrac{1}{e^2}$.

4. $x^{\frac{1}{1-x}}$, $x = 1$. Ans. $\dfrac{1}{e}$.

5. $(\log x)^{x-1}$, $x = 1$. Ans. 1.

6. $\left(1 + \dfrac{a}{x}\right)^x$, $x = \infty$. Ans. e^a.

7. $(\cot x)^{\sin x}$, $x = 0$. Ans. 1.

8. $(\sin x)^{\tan x}$, $x = \dfrac{\pi}{2}$. Ans. 1.

9. $(x-1)^{\frac{x}{\log \sin \pi x}}$, $x = 1$. Ans. e^a.

10. $\left(\tan \dfrac{\pi x}{4}\right)^{\tan \frac{\pi x}{2}}$, $x = 1$. Ans. $\dfrac{1}{e}$.

11. $\left(\tan\dfrac{\pi x}{2}\right)^{\tan \pi x}$, $\quad x = 1.$ $\quad$ Ans. $1.$

12. $\left(2 - \dfrac{x}{a}\right)^{\tan\frac{\pi x}{2a}}$, $\quad x = a.$ $\quad$ Ans. $e^{\frac{2}{\pi}}.$

13. $(\cot x)^{\frac{1}{\log x}}$, $\quad x = 0.$ $\quad$ Ans. $\dfrac{1}{e}.$

14. $[\log(e + x)]^{\frac{1}{x}}$, $\quad x = 0.$ $\quad$ Ans. $e^{\frac{1}{e}}.$

15. $(\log x)^x$, $\quad x = 0.$ $\quad$ Ans. $1.$

16. $(e^x + x)^{\frac{1}{x}}$, $\quad x = 0.$ $\quad$ Ans. $e^2.$

CHAPTER IX.

PARTIAL DIFFERENTIATION.

59. *Functions of several Independent Variables.* In the preceding chapters differentiation has been applied only to functions of a single independent variable. We shall now consider functions of two or more independent variables.

60. *Partial Differential Coefficients.* Representing by u a function of the two independent variables x and y,

$$u = f(x, y) \quad \ldots \ldots \ldots \quad (a)$$

If we differentiate (a), supposing x to vary and y to remain constant, we obtain $\dfrac{du}{dx}$.

If we differentiate (a), supposing y to vary and x to remain constant, we obtain $\dfrac{du}{dy}$.

The differential coefficients, $\dfrac{du}{dx}, \dfrac{du}{dy}$, thus derived, are called *partial differential coefficients* and are denoted by $\dfrac{\partial u}{\partial x}, \dfrac{\partial u}{\partial y}$.

For example, if $\quad u = x^3 + 3x^2y - y^3$,

$$\frac{\partial u}{\partial x} = 3x^2 + 6xy, \quad \text{regarding } y \text{ as constant.}$$

$$\frac{\partial u}{\partial y} = 3x^2 - 3y^2, \quad \text{regarding } x \text{ as constant.}$$

In general, whatever the number of independent variables, the partial differential coefficients are obtained by supposing only one to vary at a time.

EXAMPLES.

1. If $u = x^3 y^2 - 2xy^4 + 3x^2y^3$,

   show that $x\dfrac{\partial u}{\partial x} + y\dfrac{\partial u}{\partial y} = 5u$.

2. $u = (y-z)(z-x)(x-y)$, $\quad \dfrac{\partial u}{\partial x} + \dfrac{\partial u}{\partial y} + \dfrac{\partial u}{\partial z} = 0$.

3. $u = \log(x^3 + y^3 + z^3 - 3xyz)$, $\quad \dfrac{\partial u}{\partial x} + \dfrac{\partial u}{\partial y} + \dfrac{\partial u}{\partial z} = \dfrac{3}{x+y+z}$.

4. $u = \dfrac{e^{xy}}{e^x + e^y}$, $\quad \dfrac{\partial u}{\partial x} + \dfrac{\partial u}{\partial y} = (x+y-1)u$.

5. $u = \log(x + \sqrt{x^2 + y^2})$, $\quad x\dfrac{\partial u}{\partial x} + y\dfrac{\partial u}{\partial y} = 1$.

6. $u = e^x \sin y + e^y \sin x$,

   $\left(\dfrac{\partial u}{\partial x}\right)^2 + \left(\dfrac{\partial u}{\partial y}\right)^2 = e^{2x} + e^{2y} + 2e^{x+y}\sin(x+y)$.

7. $u = \log(\tan x + \tan y + \tan z)$,

   $\sin 2x \dfrac{\partial u}{\partial x} + \sin 2y \dfrac{\partial u}{\partial y} + \sin 2z \dfrac{\partial u}{\partial z} = 2$.

61. *Partial Differential Coefficients of Higher Orders.* By successive differentiation, regarding the independent variables as varying only one at a time, we may obtain

$$\dfrac{\partial^2 u}{\partial x^2}, \ \dfrac{\partial^2 u}{\partial y^2}, \ \dfrac{\partial^3 u}{\partial x^3}, \ \dfrac{\partial^4 u}{\partial y^4}, \ \ldots.$$

If we differentiate u with respect to x, then this result with respect to y, we obtain $\dfrac{\partial}{\partial y}\left(\dfrac{\partial u}{\partial x}\right)$, which is written $\dfrac{\partial^2 u}{\partial y \partial x}$.

Similarly, $\dfrac{\partial^3 u}{\partial y \partial x^2}$ is the result of three successive differentiations, two with respect to x, and one with respect to y. It will now be shown that this result is independent of the order of these differentiations.

That is, $\dfrac{\partial^2 u}{\partial y \partial x} = \dfrac{\partial^2 u}{\partial x \partial y}$, $\dfrac{\partial^3 u}{\partial y \partial x^2} = \dfrac{\partial^3 u}{\partial x \partial y \partial x} = \dfrac{\partial^3 u}{\partial x^2 \partial y}$.

62. Given $\quad u = f(x, y)$ . . . . (a)

to prove that $\quad \dfrac{\partial}{\partial y}\left(\dfrac{\partial u}{\partial x}\right) = \dfrac{\partial}{\partial x}\left(\dfrac{\partial u}{\partial y}\right)$.

Supposing x alone to change in (a),

$$\dfrac{\Delta u}{\Delta x} = \dfrac{f(x + \Delta x, y) - f(x, y)}{\Delta x} \quad . \quad . \quad . \quad . \quad (b)$$

Now supposing y alone to change in (b),

$$\dfrac{\Delta}{\Delta y}\left(\dfrac{\Delta u}{\Delta x}\right) = \dfrac{f(x+\Delta x, y+\Delta y) - f(x, y+\Delta y) - f(x+\Delta x, y) + f(x, y)}{\Delta y \, \Delta x}.$$

Reversing the above order, we find

$$\dfrac{\Delta u}{\Delta y} = \dfrac{f(x, y+\Delta y) - f(x, y)}{\Delta y},$$

and

$$\dfrac{\Delta}{\Delta x}\left(\dfrac{\Delta u}{\Delta y}\right) = \dfrac{f(x+\Delta x, y+\Delta y) - f(x+\Delta x, y) - f(x, y+\Delta y) + f(x, y)}{\Delta x \, \Delta y}.$$

Hence $\quad \dfrac{\Delta}{\Delta y}\left(\dfrac{\Delta u}{\Delta x}\right) = \dfrac{\Delta}{\Delta x}\left(\dfrac{\Delta u}{\Delta y}\right)$.

This being true, however small Δx and Δy may be, we have for the limits of the above

$$\dfrac{\partial}{\partial y}\left(\dfrac{\partial u}{\partial x}\right) = \dfrac{\partial}{\partial x}\left(\dfrac{\partial u}{\partial y}\right), \text{ or } \dfrac{\partial^2 u}{\partial y \partial x} = \dfrac{\partial^2 u}{\partial x \partial y}.$$

63. This principle, *that the order of differentiation is immaterial*, may be extended to any number of differentiations.

Thus, $\quad \dfrac{\partial^3 u}{\partial y \partial x^2} = \dfrac{\partial^2}{\partial y \partial x}\left(\dfrac{\partial u}{\partial x}\right) = \dfrac{\partial^2}{\partial x \partial y}\left(\dfrac{\partial u}{\partial x}\right) = \dfrac{\partial^3 u}{\partial x \partial y \partial x}$

$$= \dfrac{\partial}{\partial x}\left(\dfrac{\partial^2 u}{\partial y \partial x}\right) = \dfrac{\partial}{\partial x}\left(\dfrac{\partial^2 u}{\partial x \partial y}\right) = \dfrac{\partial^3 u}{\partial x^2 \partial y}.$$

It is evident that the principle applies also to functions of three or more variables.

EXAMPLES.

Verify $\dfrac{\partial^2 u}{\partial y \partial x} = \dfrac{\partial^2 u}{\partial x \partial y}$, in each of the four following equations:

1. $u = y \log(1 + xy)$.   3. $u = \sin(xy^2)$.

2. $u = x^y$.   4. $u = \dfrac{ax - by}{ay - bx}$.

5. If $u = \dfrac{x^2 y^2}{x + y}$, show that $x\dfrac{\partial^2 u}{\partial x^2} + y\dfrac{\partial^2 u}{\partial x \partial y} = 2\dfrac{\partial u}{\partial x}$.

6. $u = (x^2 + y^2)^{\frac{1}{3}}$, $3x\dfrac{\partial^2 u}{\partial x \partial y} + 3y\dfrac{\partial^2 u}{\partial y^2} + \dfrac{\partial u}{\partial y} = 0$.

7. $u = e^{xyz}$, $\dfrac{\partial^3 u}{\partial x \partial y \partial z} = (1 + 3xyz + x^2 y^2 z^2)u$.

8. $u = y^2 z^2 e^{\frac{x}{z}} + z^2 x^2 e^{\frac{y}{z}} + x^2 y^2 e^{\frac{z}{x}}$, $\dfrac{\partial^6 u}{\partial x^2 \partial y^2 \partial z^2} = e^{\frac{x}{z}} + e^{\frac{y}{z}} + e^{\frac{z}{x}}$.

9. $u = \sin(y + z)\sin(z + x)\sin(x + y)$,

$$\dfrac{\partial^3 u}{\partial x \partial y \partial z} = 2\cos(2x + 2y + 2z).$$

64. Total Differential. If u is a function of two or more variables, and all vary at the same time, the change in u is called the *total increment*, and if infinitely small, the *total differential* of u.

This total differential of u may be obtained by the usual formulae of differentiation, using differentials as in Art. 28.

For example, suppose
$$u = x^3 y - 3 x^2 y^2.$$

Differentiating, regarding both x and y variable,
$$\begin{aligned}
du &= d(x^3 y) - d(3 x^2 y^2) \\
&= x^3 dy + y\, d(x^3) - 3 x^2 d(y^2) - 3 y^2 d(x^2) \\
&= x^3 dy + 3 x^2 y\, dx - 6 x^2 y\, dy - 6 xy^2 dx \\
&= (3 x^2 y - 6 xy^2) dx + (x^3 - 6 x^2 y) dy.
\end{aligned}$$

But $\quad 3x^2y - 6xy^2 = \dfrac{\partial u}{\partial x},\quad$ and $\quad x^3 - 6x^2y = \dfrac{\partial u}{\partial y}.$

Hence $\qquad du = \dfrac{\partial u}{\partial x}dx + \dfrac{\partial u}{\partial y}dy \quad \ldots \ldots \quad (1)$

This expression for the total differential holds for *any* function of two variables, $\quad u = f(x, y)$.

For, if we differentiate this equation, using differentials as in the preceding example, we may arrange the terms in two groups containing dx and dy respectively, so that the result will be of the form

$$du = Pdx + Qdy \quad \ldots \ldots \quad (2)$$

Now if x alone varies, y being constant, (2) becomes

$$d_xu = Pdx, \quad \text{giving} \quad \dfrac{\partial u}{\partial x} = P.$$

If y alone varies, x being constant, (2) becomes

$$d_yu = Qdy, \quad \text{giving} \quad \dfrac{\partial u}{\partial y} = Q.$$

Substituting in (2) these expressions for P and Q, we have

$$du = \dfrac{\partial u}{\partial x}dx + \dfrac{\partial u}{\partial y}dy \quad \ldots \ldots \quad (3)$$

Similarly, if $\quad u = f(x, y, z),\quad$ it may be shown that

$$du = \dfrac{\partial u}{\partial x}dx + \dfrac{\partial u}{\partial y}dy + \dfrac{\partial u}{\partial z}dz \quad \ldots \ldots \quad (4)$$

65. The result of the preceding article may be reached also in the following way. *The total differential of a function of several independent variables is the sum of its partial differentials arising from the separate variation of the variables.*

Let Δu, du, denote the total increment, and differential of u. Δ_xu, Δ_yu, d_xu, d_yu, the partial increments and differentials, when x and y vary separately.

Let
$\qquad\quad u = f(x, y),$
$\qquad\quad u' = f(x + \Delta x, y),$
$\qquad\quad u'' = f(x + \Delta x, y + \Delta y).$

Then
$$\Delta_x u = u' - u,$$
$$\Delta_y u' = u'' - u',$$
$$\Delta u = u'' - u.$$

Hence
$$\Delta u = \Delta_x u + \Delta_y u'.$$

Now if Δx, Δy, and consequently $\Delta_x u$, $\Delta_y u'$, Δu, become infinitely small, we have
$$du = d_x u + d_y u,$$
since the limit of u' is u.

We may write $d_x u = \dfrac{\partial u}{\partial x} dx$, $d_y u = \dfrac{\partial u}{\partial y} dy$,

giving
$$du = \frac{\partial u}{\partial x} dx + \frac{\partial u}{\partial y} dy.$$

The process above may be extended to functions of three or more variables.

EXAMPLES.

Find as in Art. 64 the total differential of u in each of the following, and show that it agrees with (3), Art. 64.

1. $u = ax^2 + 2bxy + cy^2$, $du = 2(ax+by)dx + 2(bx+cy)dy$.

2. $u = x^{\log y}$, $\quad du = u\left(\dfrac{\log y}{x} dx + \dfrac{\log x}{y} dy\right).$

3. $u = \log \dfrac{x-y}{x+y} + 2\tan^{-1}\dfrac{x}{y}$, $\quad du = \dfrac{4x^2}{x^4 - y^4}(ydx - xdy)$.

Find the total differential of u in each of the following, and show that it agrees with (4), Art. 64.

4. $u = ax^2 + by^2 + cz^2 + 2fyz + 2gzx + 2hxy$,
$du = 2(ax+hy+gz)dx + 2(hx+by+fz)dy + 2(gx+fy+cz)dz$.

5. $u = x^{y^z}$, $\quad du = x^{y^z-1}(yzdx + zx\log xdy + xy\log xdz)$.

6. $u = \tan^2 x \tan^2 y \tan^2 z$, $du = 4u\left(\dfrac{dx}{\sin 2x} + \dfrac{dy}{\sin 2y} + \dfrac{dz}{\sin 2z}\right)$.

PARTIAL DIFFERENTIATION.

66. *Condition for an Exact Differential.*

The expression $Pdx + Qdy$ is called an *exact differential*, when it is the total differential of some function of x and y.

For example, $ydx + xdy$ is an exact differential, because it is the differential of xy.

But $2ydx + xdy$ is *not* an exact differential, because it is not the differential of *any* function of x and y.

The general expression $Pdx + Qdy$ is an exact differential only when P and Q satisfy a certain condition, which we will now derive.

Suppose this expression to be the differential of some function u, of x and y.

Then $\qquad du = Pdx + Qdy.$

But from (3) Art. 64, $\quad du = \dfrac{\partial u}{\partial x}dx + \dfrac{\partial u}{\partial y}dy.$

Hence $\qquad P = \dfrac{\partial u}{\partial x}, \quad Q = \dfrac{\partial u}{\partial y}.$

Differentiating the first of these equations with respect to y, and the second with respect to x, we have

$$\frac{\partial P}{\partial y} = \frac{\partial^2 u}{\partial y \partial x}, \quad \frac{\partial Q}{\partial x} = \frac{\partial^2 u}{\partial x \partial y}.$$

Hence $\qquad \dfrac{\partial P}{\partial y} = \dfrac{\partial Q}{\partial x} \quad \ldots \ldots \ldots$ (1)

which is the condition that $Pdx + Qdy$ may be an exact differential.

Similarly, it may be shown that $Pdx + Qdy + Rdz$ is an exact differential, when

$$\frac{\partial P}{\partial y} = \frac{\partial Q}{\partial x}, \quad \frac{\partial Q}{\partial z} = \frac{\partial R}{\partial y}, \quad \frac{\partial R}{\partial x} = \frac{\partial P}{\partial z}. \quad \ldots \ldots \text{(2)}$$

EXAMPLES.

By means of (1) determine which of the following expressions are exact differentials:

1. $(3xy + 2y^2)dx + (x^2 + 2xy)dy$.
2. $(3x^2y + 2xy^2)dx + (x^3 + 2x^2y)dy$.
3. $(xy - y^2 + 1)dx + (x^2 - xy - 1)dy$.
4. $e^{xy}[(xy - y^2 + 1)dx + (x^2 - xy - 1)dy]$.

Show that condition (1) is satisfied by the answers to Examples 1, 3, Art. 65; and conditions (2) by the answers to Examples 4, 5, Art. 65.

67. *Differentiation of an Implicit Function.* The differential coefficient of an implicit function may be expressed in terms of partial differential coefficients.

Suppose y and x connected by the equation $\phi(x, y) = 0$. Let u represent the first member of this equation. That is,

$$u = \phi(x, y) = 0 \quad \ldots \ldots \quad (1)$$

From (3) Art. 64, we have for the total differential of u,

$$du = \frac{\partial u}{\partial x}dx + \frac{\partial u}{\partial y}dy.$$

But by (1), u is always zero, that is, a constant; and therefore its total differential du must be zero. Hence

$$\frac{\partial u}{\partial x}dx + \frac{\partial u}{\partial y}dy = 0,$$

$$\therefore \frac{dy}{dx} = -\frac{\frac{\partial u}{\partial x}}{\frac{\partial u}{\partial y}}. \quad \ldots \ldots \quad (2)$$

For example, suppose, as in Art. 31,

$$a^2y^2 + b^2x^2 - a^2b^2 = 0.$$

PARTIAL DIFFERENTIATION. 87

Here $u = a^2y^2 + b^2x^2 - a^2b^2,$

and $\dfrac{\partial u}{\partial x} = 2b^2x, \quad \dfrac{\partial u}{\partial y} = 2a^2y.$

Hence by (2), $\dfrac{dy}{dx} = -\dfrac{2b^2x}{2a^2y} = -\dfrac{b^2x}{a^2y}.$

Derive by (2) the expressions for $\dfrac{dy}{dx}$ in the examples in Art. 31.

68. *Extension of Taylor's Theorem to functions of two independent variables.* If we apply Taylor's Theorem to
$$f(x+h, y+k),$$
regarding x as the only variable, we have

$$f(x+h, y+k) = f(x, y+k) + h\dfrac{\partial}{\partial x}f(x, y+k)$$
$$+ \dfrac{h^2}{\lfloor 2}\dfrac{\partial^2}{\partial x^2}f(x, y+k) + \cdots. \quad (1)$$

Now expanding $f(x, y+k)$, regarding y as the only variable,

$$f(x, y+k) = f(x, y) + k\dfrac{\partial}{\partial y}f(x, y) + \dfrac{k^2}{\lfloor 2}\dfrac{\partial^2}{\partial y^2}f(x, y) + \cdots.$$

Substituting this in (1),

$$f(x+h, y+k) = f(x, y) + h\dfrac{\partial}{\partial x}f(x, y) + k\dfrac{\partial}{\partial y}f(x, y)$$
$$+ \dfrac{1}{\lfloor 2}\left[h^2\dfrac{\partial^2}{\partial x^2}f(x, y) + 2hk\dfrac{\partial^2}{\partial x\partial y}f(x, y) + k^2\dfrac{\partial^2}{\partial y^2}f(x, y)\right] + \cdots.$$

This may be expressed in the symbolic form thus:

$$f(x+h, y+k) = f(x, y) + \left(h\dfrac{\partial}{\partial x} + k\dfrac{\partial}{\partial y}\right)f(x, y)$$
$$+ \dfrac{1}{\lfloor 2}\left(h\dfrac{\partial}{\partial x} + k\dfrac{\partial}{\partial y}\right)^2 f(x, y) + \dfrac{1}{\lfloor 3}\left(h\dfrac{\partial}{\partial x} + k\dfrac{\partial}{\partial y}\right)^3 f(x, y) + \cdots$$

where $\left(h\dfrac{\partial}{\partial x}+k\dfrac{\partial}{\partial y}\right)^n$ is to be expanded by the Binomial Theorem, as if $h\dfrac{\partial}{\partial x}$ and $k\dfrac{\partial}{\partial y}$ were the two terms of the binomial, and the resulting terms applied separately to $f(x, y)$.

69. *Taylor's Theorem applied to functions of any number of independent variables.* By a method similar to that of the preceding article we shall find

$$f(x+h,\ y+k,\ z+l) = f(x,\ y,\ z) + \left(h\dfrac{\partial}{\partial x} + k\dfrac{\partial}{\partial y} + l\dfrac{\partial}{\partial z}\right)f(x,\ y,\ z)$$
$$+ \dfrac{1}{\lfloor 2}\left(h\dfrac{\partial}{\partial x} + k\dfrac{\partial}{\partial y} + l\dfrac{\partial}{\partial z}\right)^2 f(x,\ y,\ z) + \cdots.$$

This expansion may be extended to any number of variables.

CHAPTER X.

CHANGE OF THE VARIABLES IN DIFFERENTIAL COEFFICIENTS.

70. *To express* $\dfrac{dy}{dx}, \dfrac{d^2y}{dx^2}, \dfrac{d^3y}{dx^3}, \ldots$ *in terms of* $\dfrac{dx}{dy}, \dfrac{d^2x}{dy^2}, \dfrac{d^3x}{dy^3}, \ldots.$

This is called *changing the independent variable from x to y.*

By (1) Art. 21, $\quad \dfrac{dy}{dx} = \dfrac{1}{\dfrac{dx}{dy}} \quad \cdots \cdots \cdots \cdots \cdots$ (a)

By (1) Art. 22, $\quad \dfrac{d^2y}{dx^2} = \dfrac{d}{dx}\dfrac{dy}{dx} = \dfrac{d}{dy}\dfrac{dy}{dx} \cdot \dfrac{dy}{dx}.$

From (a), $\quad \dfrac{d}{dy}\dfrac{dy}{dx} = \dfrac{d}{dy}\dfrac{1}{\dfrac{dx}{dy}} = -\dfrac{\dfrac{d^2x}{dy^2}}{\left(\dfrac{dx}{dy}\right)^2}.$

$\therefore \dfrac{d^2y}{dx^2} = -\dfrac{\dfrac{d^2x}{dy^2}}{\left(\dfrac{dx}{dy}\right)^3} \quad \cdots \cdots \cdots \cdots$ (b)

Similarly, $\quad \dfrac{d^3y}{dx^3} = \dfrac{d}{dx}\dfrac{d^2y}{dx^2} = \dfrac{d}{dy}\dfrac{d^2y}{dx^2} \cdot \dfrac{dy}{dx}.$

From (b), $\quad \dfrac{d}{dy}\dfrac{d^2y}{dx^2} = \dfrac{3\left(\dfrac{d^2x}{dy^2}\right)^2 - \dfrac{dx}{dy}\dfrac{d^3x}{dy^3}}{\left(\dfrac{dx}{dy}\right)^4}.$

$\therefore \dfrac{d^3y}{dx^3} = \dfrac{3\left(\dfrac{d^2x}{dy^2}\right)^2 - \dfrac{dx}{dy}\dfrac{d^3x}{dy^3}}{\left(\dfrac{dx}{dy}\right)^5} \quad \cdots \cdots$ (c)

71. It is sometimes necessary in the differential coefficients,

$$\frac{dy}{dx},\ \frac{d^2y}{dx^2},\ \frac{d^3y}{dx^3},\ \ldots,$$

to introduce a new variable z in place of x or y, z being a given function of the variable removed.

There are two cases, according as z replaces y or x.

72. *First.* To express $\frac{dy}{dx},\ \frac{d^2y}{dx^2},\ \frac{d^3y}{dx^3},\ \ldots$ in terms of $\frac{dz}{dx}$, $\frac{d^2z}{dx^2},\ \frac{d^3z}{dx^3},\ \ldots,$ where y is a given function of z.

For example, suppose $\quad y = z^3.$

Then $\quad \dfrac{dy}{dx} = 3z^2 \dfrac{dz}{dx}.$

$$\frac{d^2y}{dx^2} = 6z\left(\frac{dz}{dx}\right)^2 + 3z^2 \frac{d^2z}{dx^2}.$$

$$\frac{d^3y}{dx^3} = 6\left(\frac{dz}{dx}\right)^3 + 18z\frac{dz}{dx}\frac{d^2z}{dx^2} + 3z^2\frac{d^3z}{dx^3}.$$

Similarly, $\dfrac{d^4y}{dx^4},\ \dfrac{d^5y}{dx^5},\ \ldots,$ may be expressed in terms of z and x.

It is to be noticed that in this case there is no change of the *independent* variable, which remains x.

73. *Second.* To express $\dfrac{dy}{dx},\ \dfrac{d^2y}{dx^2},\ \dfrac{d^3y}{dx^3},\ \ldots$ in terms of $\dfrac{dy}{dz}$, $\dfrac{d^2y}{dz^2},\ \dfrac{d^3y}{dz^3},\ \ldots,$ where x is a given function of z.

This is called *changing the independent variable from x to z.*

For example, suppose $\quad x = z^3.$

By (1) Art. 22, $\quad \dfrac{dy}{dx} = \dfrac{dy}{dz}\dfrac{dz}{dx}.$

But $\dfrac{dx}{dz} = 3z^2$, $\dfrac{dz}{dx} = \dfrac{z^{-2}}{3}$.

$$\therefore \dfrac{dy}{dx} = \dfrac{1}{3} z^{-2} \dfrac{dy}{dz} \ . \ . \ . \ . \ . \ . \ . \ . \ . \ (a)$$

By (1) Art. 22, $\dfrac{d^2y}{dx^2} = \dfrac{d}{dx}\dfrac{dy}{dx} = \dfrac{d}{dz}\dfrac{dy}{dx} \cdot \dfrac{dz}{dx}$.

From (a), $\quad \dfrac{d}{dz}\dfrac{dy}{dx} = \dfrac{1}{3}\left(z^{-2}\dfrac{d^2y}{dz^2} - 2z^{-3}\dfrac{dy}{dz}\right)$.

$$\therefore \dfrac{d^2y}{dx^2} = \dfrac{1}{9}\left(z^{-4}\dfrac{d^2y}{dz^2} - 2z^{-5}\dfrac{dy}{dz}\right). \ . \ . \ . \ . \ . \ (b)$$

Similarly, $\quad \dfrac{d^3y}{dx^3} = \dfrac{d}{dz}\dfrac{d^2y}{dx^2} \cdot \dfrac{dz}{dx}$.

From (b), $\quad \dfrac{d}{dz}\dfrac{d^2y}{dx^2} = \dfrac{1}{9}\left(z^{-4}\dfrac{d^3y}{dz^3} - 6z^{-5}\dfrac{d^2y}{dz^2} + 10z^{-6}\dfrac{dy}{dz}\right)$.

$$\therefore \dfrac{d^3y}{dx^3} = \dfrac{1}{27}\left(z^{-6}\dfrac{d^3y}{dz^3} - 6z^{-7}\dfrac{d^2y}{dz^2} + 10z^{-8}\dfrac{dy}{dz}\right).$$

EXAMPLES.

Change the independent variable from x to y in the two following equations:

1. $3\left(\dfrac{d^2y}{dx^2}\right)^2 - \dfrac{dy}{dx}\dfrac{d^3y}{dx^3} - \dfrac{d^2y}{dx^2}\left(\dfrac{dy}{dx}\right)^2 = 0.$ *Ans.* $\dfrac{d^3x}{dy^3} + \dfrac{d^2x}{dy^2} = 0.$

2. $\left(3a\dfrac{dy}{dx} + 2\right)\left(\dfrac{d^2y}{dx^2}\right)^2 = \left(a\dfrac{dy}{dx} + 1\right)\dfrac{dy}{dx}\dfrac{d^3y}{dx^3}.$

$$\textit{Ans.} \ \left(\dfrac{d^2x}{dy^2}\right)^2 = \left(\dfrac{dx}{dy} + a\right)\dfrac{d^3x}{dy^3}.$$

Change the variable from y to z in the two following equations:

3. $\dfrac{d^2y}{dx^2} = 1 + \dfrac{2(1+y)}{1+y^2}\left(\dfrac{dy}{dx}\right)^2$, $\quad y = \tan z$.

$\quad\quad\quad\quad$ Ans. $\dfrac{d^2z}{dx^2} - 2\left(\dfrac{dz}{dx}\right)^2 = \cos^2 z$.

4. $(1+y)^2\left(\dfrac{d^3y}{dx^3} - 2y\right) + \left(\dfrac{dy}{dx}\right)^3 = 2(1+y)\dfrac{dy}{dx}\dfrac{d^2y}{dx^2}$, $\quad y = z^2 + 2z$.

$\quad\quad\quad\quad$ Ans. $(z+1)\dfrac{d^3z}{dx^3} = \dfrac{dz}{dx}\dfrac{d^2z}{dx^2} + z^2 + 2z$.

Change the independent variable from x to z in the following equations:

5. $\dfrac{d^2y}{dx^2} + \dfrac{1}{x}\dfrac{dy}{dx} + y = 0$, $\quad x^2 = 4z$. $\quad$ Ans. $z\dfrac{d^2y}{dz^2} + \dfrac{dy}{dz} + y = 0$.

6. $\dfrac{d^2y}{dx^2} + \dfrac{2x}{1+x^2}\dfrac{dy}{dx} + \dfrac{y}{(1+x^2)^2} = 0$, $\quad x = \tan z$.

$\quad\quad\quad\quad$ Ans. $\dfrac{d^2y}{dz^2} + y = 0$.

7. $(2x-1)^3\dfrac{d^3y}{dx^3} + (2x-1)\dfrac{dy}{dx} = 2y$, $\quad 2x = 1 + e^z$.

$\quad\quad\quad\quad$ Ans. $4\dfrac{d^3y}{dz^3} - 12\dfrac{d^2y}{dz^2} + 9\dfrac{dy}{dz} - y = 0$.

8. $x^4\dfrac{d^4y}{dx^4} + 6x^3\dfrac{d^3y}{dx^3} + 9x^2\dfrac{d^2y}{dx^2} + 3x\dfrac{dy}{dx} + y = \log x$, $\quad x = e^z$.

$\quad\quad\quad\quad$ Ans. $\dfrac{d^4y}{dz^4} + 2\dfrac{d^2y}{dz^2} + y = z$.

APPLICATION TO PLANE CURVES.

CHAPTER XI.

CERTAIN CURVES IN THE FOLLOWING CHAPTERS.

74. We give in this chapter representations and descriptions of some of the curves used as examples in the following chapters.

RECTANGULAR CO-ORDINATES.

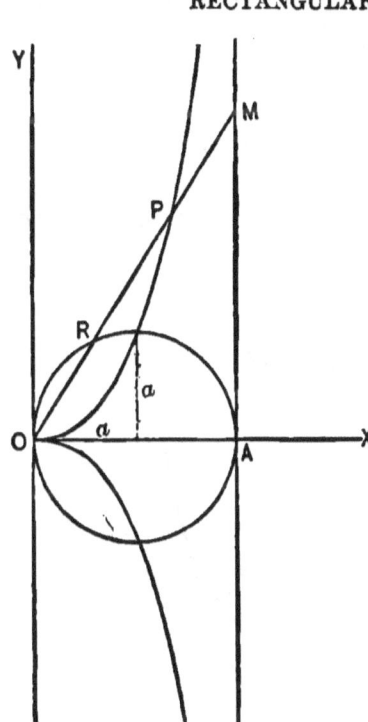

75. *The Cissoid*,

$$y^2 = \frac{x^3}{2a - x}.$$

This curve may be constructed from the circle ORA (radius, a) by drawing any oblique line OM, and making

$$MP = OR.$$

The equation above may be easily obtained from this construction. The line AM parallel to OY is an asymptote.

76. The Witch, $\quad y = \dfrac{8a^3}{x^2 + 4a^2}.$

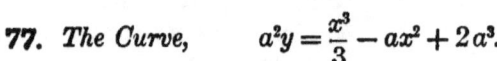

This curve may be constructed from the circle ORA (radius, a) by drawing any abscissa MR, and extending it to P by the contruction shown in the figure.

The equation above may be derived from this construction. The axis of X is an asymptote.

77. The Curve, $\quad a^2 y = \dfrac{x^3}{3} - ax^2 + 2a^3.$

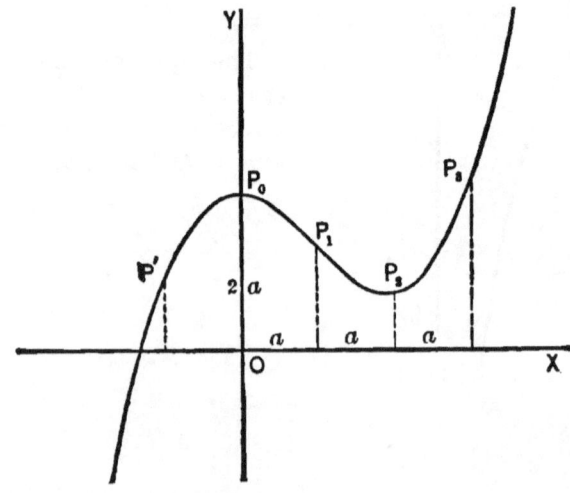

78. *The Catenary,*

$$y = \frac{a}{2}(e^{\frac{x}{a}} + e^{-\frac{x}{a}}).$$

This is the curve of a cord or chain suspended freely between two points.

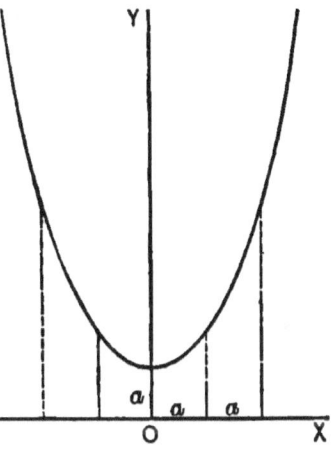

79. *The Parabola referred to Tangents at the Extremities of the Latus Rectum,* $\quad x^{\frac{1}{2}} + y^{\frac{1}{2}} = a^{\frac{1}{2}}.$

$$OL = OL' = a.$$

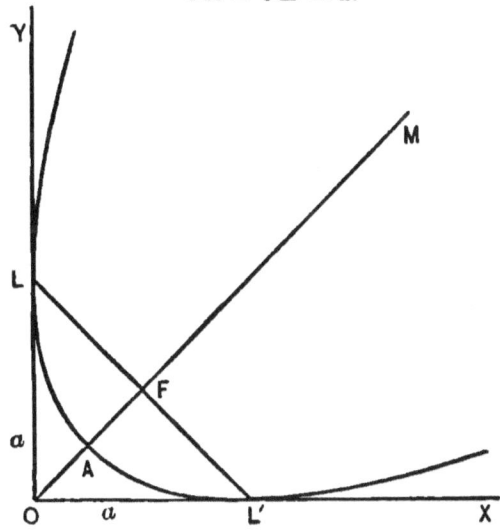

The line LL' is the latus rectum; its middle point F, the focus; OFM is the axis of the parabola; the middle **point of** OF, A, is the vertex.

80. *The Hypocycloid of Four Cusps,* $x^{\frac{2}{3}} + y^{\frac{2}{3}} = a^{\frac{2}{3}}$.

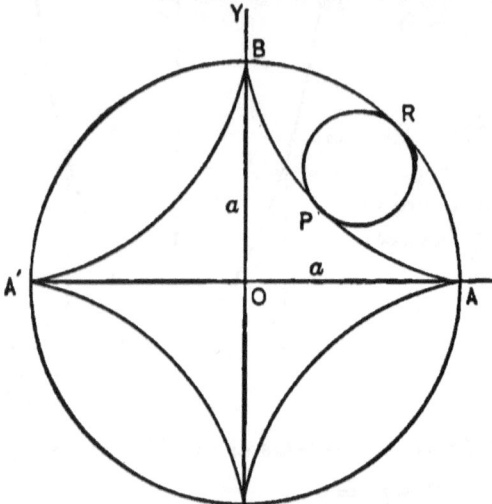

This is the curve described by a point P in the circumference of the circle PR, as it rolls within the circumference of the fixed circle ABA', whose radius, a, is four times that of the former.

81. *The Curve,* $\left(\dfrac{x}{a}\right)^{2} + \left(\dfrac{y}{b}\right)^{\frac{2}{3}} = 1.$

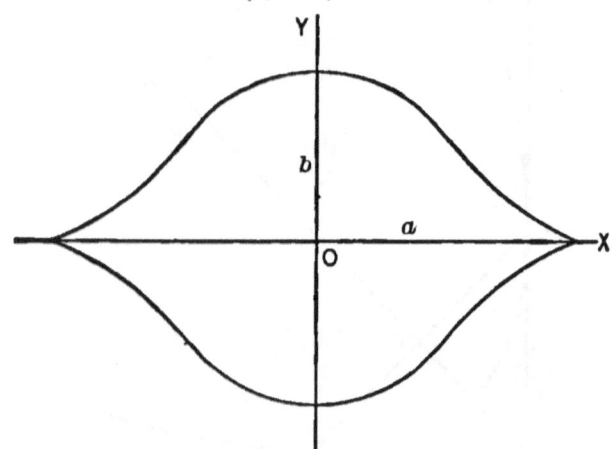

The equation is that of the ellipse

$$\left(\dfrac{x}{a}\right)^{2} + \left(\dfrac{y}{b}\right)^{2} = 1,$$

with the second exponent changed from 2 to $\frac{2}{3}$.

82. *The Curve,* $\quad a^4y^2 = a^2x^4 - x^6.$

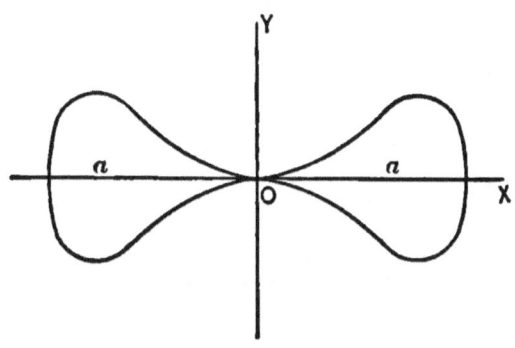

83. *The Curve,* $\quad a^{n-1}y = x^n,$ . . . . (1)

where one co-ordinate is proportional to the nth power of the other, is frequently called the *parabola of the nth degree.*

84. If $n = 3$ in (1) Art. 83, we have

The Cubical Parabola, $\quad a^2y = x^3.$

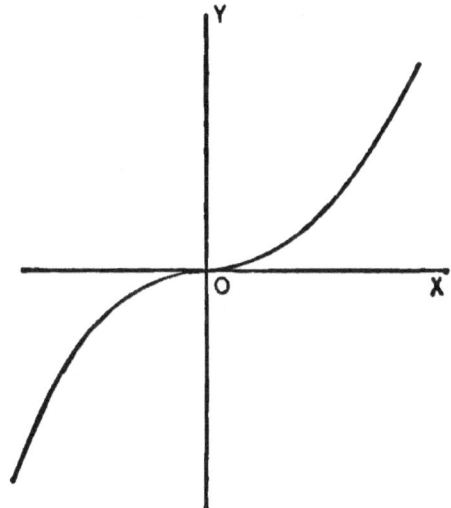

85. If $n = \frac{3}{2}$, in (1) Art. 83, we have

The Semi-Cubical Parabola, $\qquad a^{\frac{1}{2}}y = x^{\frac{3}{2}}$, or $ay^2 = x^3$.

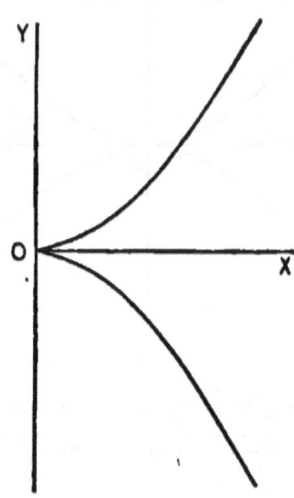

POLAR CO-ORDINATES.

86. *The Circle,* $\qquad r = a \sin \theta$.

The circle is OPA (diameter, a) tangent to the initial line OX at the pole, O.

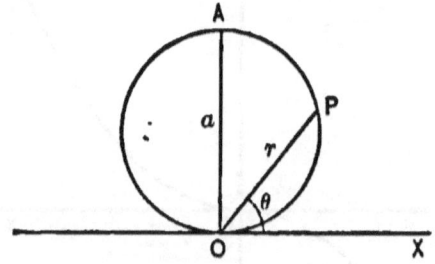

87. *The Spiral of Archimedes,* $\qquad r = a\theta$.

In this curve r is proportional to θ. Assuming $r = OA$, when $\theta = 2\pi$, then

$$OP_1 = \tfrac{1}{8}OA, \quad OP_2 = \tfrac{1}{4}OA, \quad OP_3 = \tfrac{3}{8}OA, \quad OP_4 = \tfrac{1}{2}OA, \ \ldots.$$

The dotted part of the curve corresponds to *negative* values of θ.

88. *The Logarithmic Spiral,* $r = e^{a\theta}$.

Starting from A, where $\theta = 0$ and $r = 1$, r increases with θ; but if we suppose θ negative, r decreases as θ numerically increases. Since $r = 0$ only when $\theta = -\infty$, it follows that an infinite number of retrograde revolutions from A is required to reach the pole O.

A property of this spiral is that the radii vectores OP, OP_1, OP_2, ..., make a constant angle with the curve.

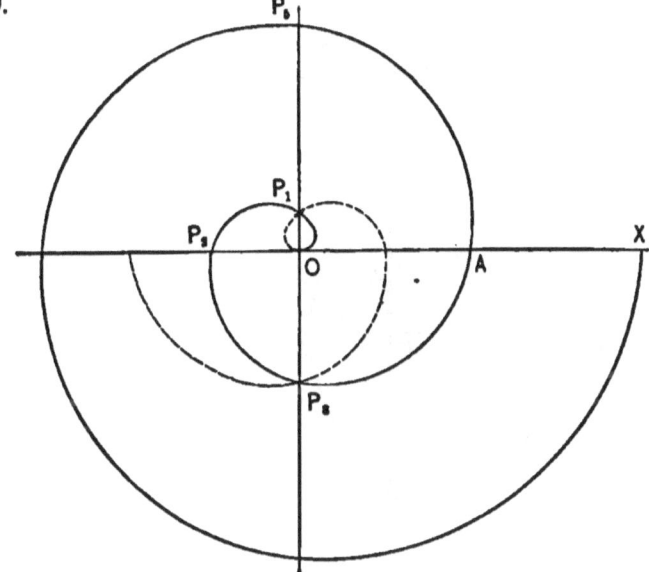

89. *The Parabola*, $\quad r = a \sec^2 \dfrac{\theta}{2}.$

The initial line OX is the axis of the parabola; the pole O is the focus; LL', the latus rectum.

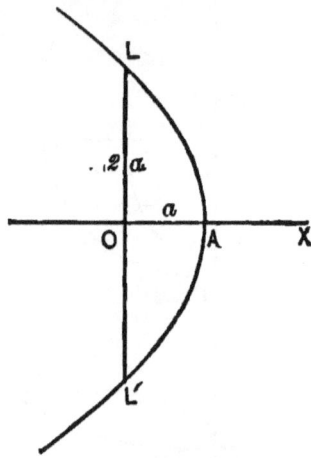

90. *The Lemniscate*, $\quad r^2 = a^2 \cos 2\theta.$

This is a curve of two loops like the figure eight.

It may be defined in connection with the equilateral hyperbola, as the locus of P, the foot of a perpendicular from O on PQ, any tangent to the hyperbola.

The loops are limited by the asymptotes of the hyperbola, making

$$TOX = T'OX = 45°. \qquad OA = a.$$

The lemniscate has the following property:

If two points, F and F', be taken on the axis, such that

$$OF = OF' = \frac{a}{\sqrt{2}},$$

then the product of the distances $P'F$, $P'F'$, of any point of the curve from these fixed points, is constant, and equal to the square of OF.

The points F and F' are called the foci of the lemniscate, and this property may be used as a definition of the curve.

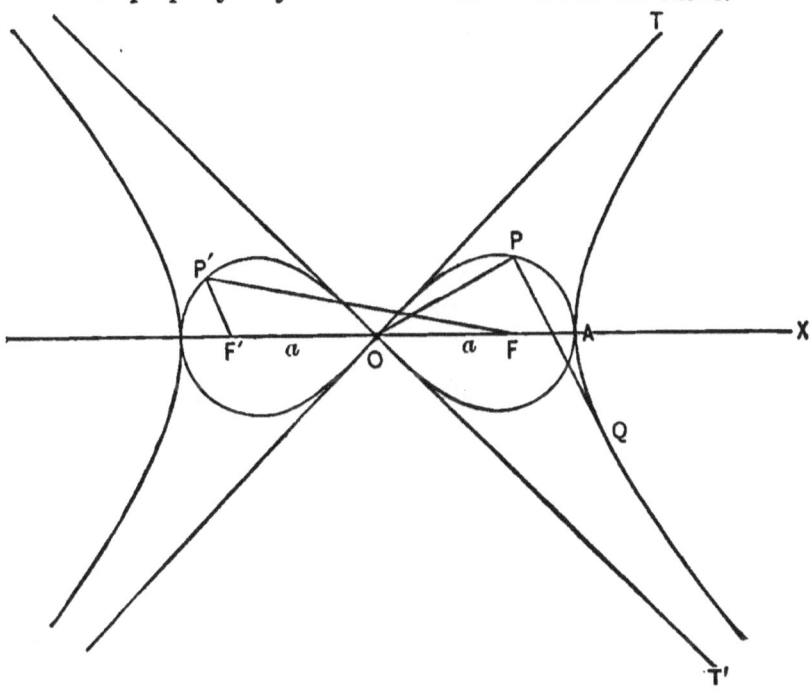

91. *The Curve,* $r = a \sin^3 \dfrac{\theta}{3}.$

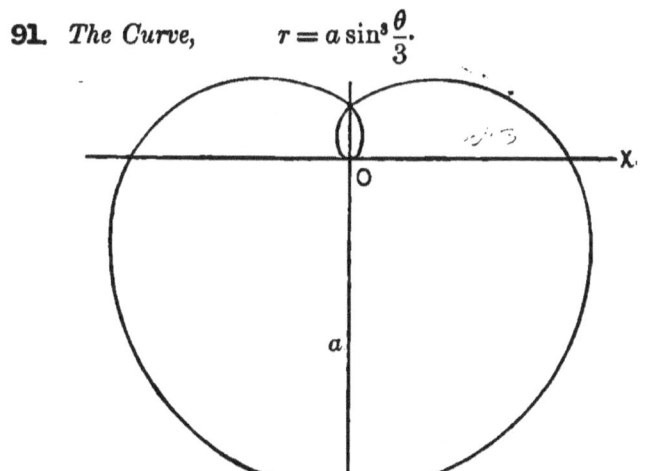

92. The Cardioid, $r = a(1 - \cos \theta)$.

This is the curve described by a point P in the circumference of a circle PA (diameter, a) as it rolls upon an equal fixed circle OA.

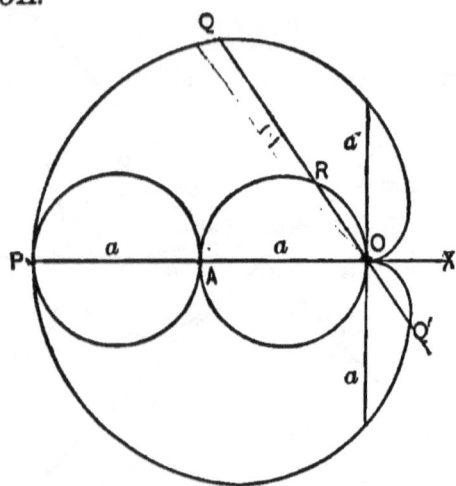

Or it may be constructed by drawing through O, any line OR in the circle OA, and producing OR to Q and Q', making $RQ = RQ' = OA$.

The given equation follows directly from this construction.

93. The Curve, $r = a \sin 2\theta$.

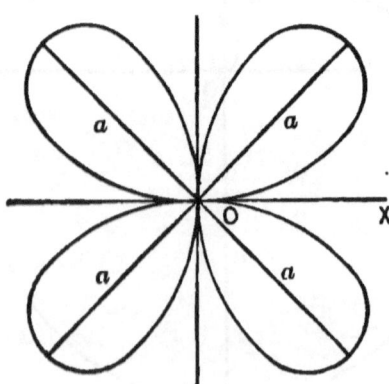

CHAPTER XII.

DIRECTION OF CURVE. TANGENT AND NORMAL. ASYMPTOTES.

94. *Direction of Curve.* When the equation of the curve is given in rectangular co-ordinates, its direction at any point is determined by the angle made by its tangent at that point with the axis of X. We shall denote this angle by ϕ.

Let P be a point in a curve whose equation is $y = f(x)$, its co-ordinates being $x = OM$, and $y = PM$. Draw the tangent PT, and PR parallel to OX. Then $TPR = \phi$.

Now give to x the increment $\Delta x = MN$; then y will receive the increment $\Delta y = QR$, and we have another point Q in the curve. Draw PQ.

Then $$\tan QPR = \frac{QR}{PR} = \frac{\Delta y}{\Delta x} \quad \cdots \cdots \quad (a)$$

Now if Δx be supposed to diminish and approach zero, Δy will approach zero, the point Q will move along the curve towards P, and PQ will approach in direction PT as its limit.

Taking the limits of the two members of equation (a), we have
$$\text{limit of } QPR = TPR = \phi,$$

and $$\text{limit of } \frac{\Delta y}{\Delta x} = \frac{dy}{dx}, \text{ by definition.}$$

$$\therefore \tan \phi = \frac{dy}{dx} \quad \cdots \cdots \cdots \quad (1)$$

104   DIFFERENTIAL CALCULUS.

For example, find the direction at any point of the parabola
$$y^2 = 4ax.$$

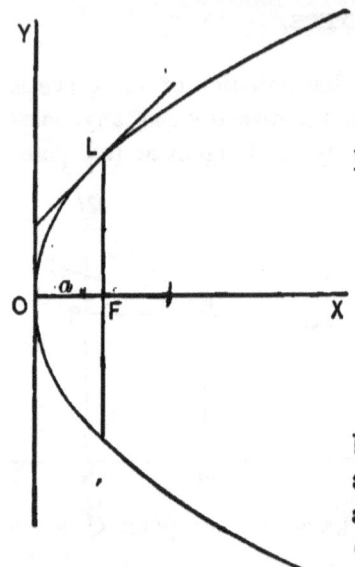

Here $\dfrac{dy}{dx} = \sqrt{\dfrac{a}{x}};$

hence $\tan \phi = \sqrt{\dfrac{a}{x}}.$

At the vertex O, where $x = 0$,
$$\tan \phi = \infty, \quad \phi = 90°.$$

At L, where $x = a$,
$$\tan \phi = 1, \quad \phi = 45°.$$

For that part of the curve beyond L, as x increases, $\tan \phi$ and ϕ decrease. Thus the parabola is more nearly parallel to OX, the further it extends from O.

95. Subtangent and Subnormal. Let PT be the tangent, and PN the normal, to a curve at the point P, whose ordinate is $y = PM$. Then MT is called the *subtangent*, and MN the *subnormal*, corresponding to the point P.

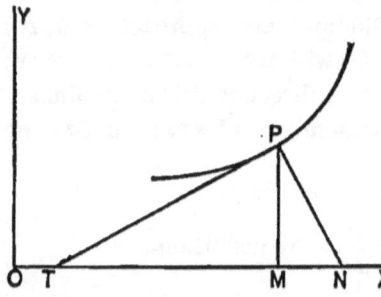

To find expressions for these quantities:

$$\text{Subtangent} = MT = PM \cot PTM = y \cot \phi = \frac{y}{\dfrac{dy}{dx}} = y\frac{dx}{dy}.$$

$$\text{Subnormal} = MN = PM \tan MPN = y \tan \phi = y\frac{dy}{dx}.$$

The length PN is sometimes called the *normal*. It is evident that
$$PN = PM \sec \phi = y\sqrt{1 + \left(\frac{dy}{dx}\right)^2}.$$

EXAMPLES.

1. The equation of a curve is $a^2y = \dfrac{x^3}{3} - ax^2 + 2a^3$.

   (a). Find ϕ when $x = 0$ and $x = a$. *Ans.* $\phi = 0$ and $135°$.

   (b). Find the points where the curve is parallel to X.

   Ans. $x = 0$ and $x = 2a$.

   (c). Find the points where $\phi = 45°$. *Ans.* $x = (1 \pm \sqrt{2})a$.

   (d). Find the point where the direction is the same as that at $x = 3a$. *Ans.* $x = -a$.

2. Where is the curve $y(x-1)(x-2) = x-3$ parallel to X? *Ans.* $x = 3 \pm \sqrt{2}$.

3. Show that the ellipse $\dfrac{x^2}{18} + \dfrac{y^2}{8} = 1$, and the hyperbola $x^2 = y^2 + 5$, intersect at right angles.

4. At what angle does the circle $x^2 + y^2 = 8ax$ intersect the cissoid $y^2 = \dfrac{x^3}{2a-x}$?

   Ans. At the origin, $90°$; at the two other points, $45°$.

5. At what angle does the parabola $x^2 = 4ay$ intersect the witch $y = \dfrac{8a^3}{x^2 + 4a^2}$? *Ans.* $\tan^{-1} 3 = 71° 33' 54''$.

6. Find the subtangent and subnormal of the parabola $y^2 = 4ax$. *Ans.* $2x$ and $2a$.

7. Find the subtangent and subnormal of the parabola of the nth degree $y^n = a^{n-1}x$. *Ans.* nx and $\dfrac{y^2}{nx}$.

8. Find the subtangent of the cissoid $y^2 = \dfrac{x^3}{2a-x}$.

   Ans. $\dfrac{2ax - x^2}{3a - x}$.

9. Find the normal of the catenary $y = \dfrac{a}{2}(e^{\frac{x}{a}} + e^{-\frac{x}{a}})$. *Ans.* $\dfrac{y^2}{a}$.

96. *Direction of Curve. Polar Co-ordinates.* By means of the equations

$$x = r\cos\theta, \quad y = r\sin\theta,$$

we may express $\tan\phi$ in terms of r and θ. Thus

$$\tan\phi = \frac{dy}{dx} = \frac{\frac{dy}{d\theta}}{\frac{dx}{d\theta}} = \frac{r\cos\theta + \frac{dr}{d\theta}\sin\theta}{-r\sin\theta + \frac{dr}{d\theta}\cos\theta} \quad \cdots \quad (a)$$

The angle OPT between the tangent and the radius vector may also be expressed. Denote this angle by ψ. Let r, θ, be the co-ordinates of P; $r + \Delta r$, $\theta + \Delta\theta$, the co-ordinates of Q. Describe the arc PR about O as a centre. Then

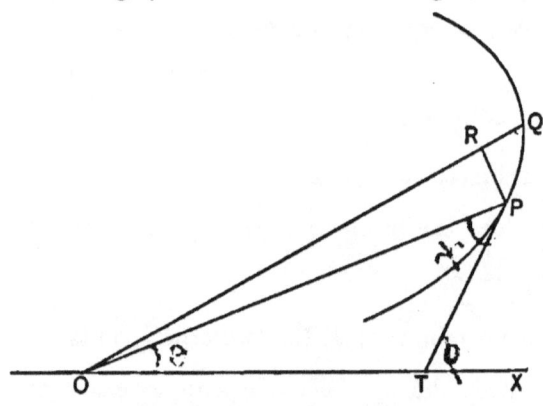

$$RQ = \Delta r, \quad POR = \Delta\theta, \quad PR = r\Delta\theta.$$

If we suppose Q to approach P, the figure PRQ will approach more and more nearly a right triangle, R being the right angle. We have at the limit

$$\tan PQR = \frac{RP}{RQ} = \frac{r\Delta\theta}{\Delta r},$$

or
$$\tan\psi = \frac{r\,d\theta}{dr} = \frac{r}{\frac{dr}{d\theta}} \quad \cdots\cdots\cdots (b)$$

We also have
$$PTX = OPT + POX,$$

or
$$\phi = \psi + \theta \quad \cdots\cdots\cdots\cdots (c)$$

97. *Polar Subtangent and Subnormal.*

If through O, NT be drawn perpendicular to OP, OT is called the *polar subtangent*, and ON the *polar subnormal*, corresponding to the point P.

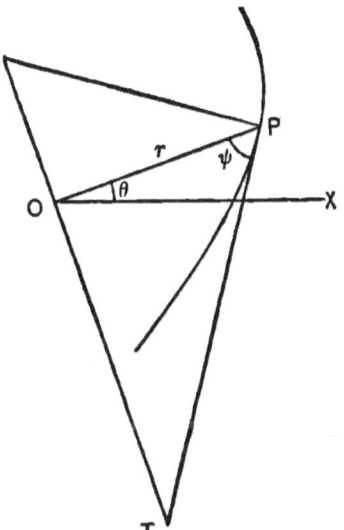

$OT = OP \tan OPT$; that is,

Polar subtangent $= r \tan \psi = \dfrac{r^2}{\dfrac{dr}{d\theta}}$.

$ON = OP \cot PNO$; that is,

Polar subnormal $= r \cot \psi = \dfrac{dr}{d\theta}$.

EXAMPLES.

1. In the circle $r = a \sin \theta$, find ψ and ϕ.
   Ans. $\psi = \theta$, and $\phi = 2\theta$.

2. In the logarithmic spiral $r = e^{a\theta}$, show that ψ is constant.

3. In the spiral of Archimedes, $r = a\theta$, show that $\tan \psi = \theta$; thence find the values of ψ when $\theta = 2\pi$ and 4π.
   Ans. $80° 57'$ and $85° 27'$.

   Also show that the polar subnormal is constant.

4. The equation of the lemniscate referred to a tangent at its centre is $r^2 = a^2 \sin 2\theta$. Find ψ, ϕ, and the polar subtangent.
   Ans. $\psi = 2\theta$; $\phi = 3\theta$; subtangent $= a \tan 2\theta \sqrt{\sin 2\theta}$.

5. Given the equation of a curve $r = a \sin^3 \dfrac{\theta}{3}$; show that $\phi = 4\psi$.

6. In the parabola $r = a \sec^2 \dfrac{\theta}{2}$, show that $\phi + \psi = \pi$.

7. In the cardioid $r = a(1 - \cos \theta)$, find ϕ, ψ, and the polar subtangent.

Ans. $\phi = \dfrac{3\theta}{2}$; $\psi = \dfrac{\theta}{2}$; subtangent $= 2\,a \tan \dfrac{\theta}{2} \sin^2 \dfrac{\theta}{2}$.

8. Find the area of the circumscribed square of the preceding cardioid, formed by tangents inclined $45°$ to the axis.

Ans. $\dfrac{27}{16}(2 + \sqrt{3})a^2$.

9. Derive equation (a) from equations (b) and (c), of Art. 96.

98. *Differential Coefficient of the Arc. Rectangular Co-ordinates.* In the figure of Art. 94, let s denote the length of the arc of the curve measured from any fixed point of it.

Then $\qquad s = \text{arc } AP, \quad \Delta s = \text{arc } PQ.$

We have $\qquad \sec QPR = \dfrac{PQ}{PR}.$

Now suppose Δx to approach zero, and the point Q to approach P.

Then $\quad \text{limit } \sec QPR = \sec TPR = \sec \phi.$

$$\text{limit} \frac{PQ}{PR} = \text{limit} \frac{\text{arc } PQ}{PR} = \text{limit} \frac{\Delta s}{\Delta x} = \frac{ds}{dx}.$$

Hence $\quad \sec \phi = \dfrac{ds}{dx}$;

therefore $\dfrac{ds}{dx} = \sqrt{1 + \tan^2 \phi} = \sqrt{1 + \left(\dfrac{dy}{dx}\right)^2}.$ . . . . (1)

It is evident also that

$$\sin \phi = \frac{dy}{ds}, \quad \cos \phi = \frac{dx}{ds}. \quad \ldots \ldots \quad (2)$$

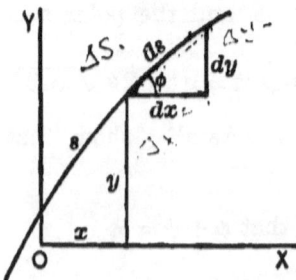

It may be noticed that these relations (1) and (2) are correctly represented by a right triangle, whose hypothenuse is ds, sides dx and dy, and angle at the base ϕ.

Here $ds = \sqrt{(dx)^2 + (dy)^2}$,

or $\qquad \dfrac{ds}{dx} = \sqrt{1 + \left(\dfrac{dy}{dx}\right)^2}.$

98½. *Differential Coefficient of the Arc. Polar Co-ordinates.* From the figure of Art. 96, by considering the limiting triangle of PRQ, we have.

$$\text{limit sec } PQR = \text{limit} \frac{PQ}{RQ} = \text{limit} \frac{\Delta s}{\Delta r},$$

or $\qquad \sec \psi = \dfrac{ds}{dr}.$ . . . . (1)

Hence $\qquad \dfrac{ds}{dr} = \sqrt{1 + \tan^2 \psi} = \sqrt{1 + r^2 \left(\dfrac{d\theta}{dr}\right)^2},$ . . . . (2)

$$\dfrac{ds}{d\theta} = \dfrac{ds}{dr}\dfrac{dr}{d\theta} = \sqrt{r^2 + \left(\dfrac{dr}{d\theta}\right)^2}. \qquad \cdots \qquad (3)$$

It may be noticed that these relations (1), (2), and (3), are correctly represented by a right triangle, whose hypothenuse is ds, sides dr and $rd\theta$, and angle between dr and ds, ψ.

Here
$$ds = \sqrt{(dr)^2 + (rd\theta)^2},$$
and thence
$$\dfrac{ds}{dr} = \sqrt{1 + r^2\left(\dfrac{d\theta}{dr}\right)^2}, \quad \text{or} \quad \dfrac{ds}{d\theta} = \sqrt{r^2 + \left(\dfrac{dr}{d\theta}\right)^2}.$$

99. *Equations of the Tangent and Normal.* Having given the equation of a curve $y = f(x)$, let it be required to find the equation of a straight line tangent to it at a given point.

Let (x', y') be the given point of contact. Then the equation of a straight line through this point is

$$y - y' = m(x - x'), \qquad \cdots \cdots \cdots \cdots (a)$$

in which x and y are the variable co-ordinates of any point in the straight line; and m, the tangent of its inclination to the axis of X. But since the line is to be tangent to the given curve, we must have, by (1) Art. 94,

$$m = \tan \phi = \dfrac{dy}{dx},$$

$\frac{dy}{dx}$ being derived from the equation of the given curve $y = f(x)$, and applied to the point of contact (x', y').

If we denote this by $\frac{dy'}{dx'}$, we have, substituting $m = \frac{dy'}{dx'}$ in equation (a),

$$y - y' = \frac{dy'}{dx'}(x - x') \quad \ldots \quad \ldots \quad (1)$$

for the equation of the required tangent.

Since the normal is a line through (x', y') perpendicular to the tangent, we have for its equation

$$y - y' = -\frac{1}{\frac{dy'}{dx'}}(x - x') = -\frac{dx'}{dy'}(x - x'). \quad \ldots \quad (2)$$

For example, find the equations of the tangent and normal to the circle $x^2 + y^2 = a^2$ at the point (x', y').

Here, by differentiating $x^2 + y^2 = a^2$, we find

$$\frac{dy}{dx} = -\frac{x}{y}, \text{ from which } \frac{dy'}{dx'} = -\frac{x'}{y'}.$$

Substituting in (1), we have

$$y - y' = -\frac{x'}{y'}(x - x'),$$

as the equation of the required tangent.

It may be simplified as follows:—

$$yy' - y'^2 = -xx' + x'^2,$$
$$xx' + yy' = x'^2 + y'^2 = a^2.$$

The equation of the normal to the circle is found from (2) to be

$$y - y' = \frac{y'}{x'}(x - x'),$$

which reduces to

$$y = \frac{y'}{x'}x.$$

EXAMPLES.

Find the equations of the tangent and normal to each of the three following curves at the point (x', y') :

1. The parabola $y^2 = 4ax$.

    Ans. $yy' = 2a(x+x')$, $2a(y-y') + y'(x-x') = 0$.

2. The ellipse $\dfrac{x^2}{a^2} + \dfrac{y^2}{b^2} = 1$.

    Ans. $\dfrac{xx'}{a^2} + \dfrac{yy'}{b^2} = 1$, $b^2 x'(y-y') = a^2 y'(x-x')$.

3. The equilateral hyperbola $2xy = a^2$.

    Ans. $xy' + yx' = a^2$, $y'(y-y') = x'(x-x')$.

4. Show that in the preceding curve the area of the triangle formed by a tangent and the co-ordinate axes is constant and equal to a^2.

5. In the cissoid $y^2 = \dfrac{x^3}{2a-x}$, find the equations of the tangent and normal at the points whose abscissa is a.

    Ans. At (a, a), $\quad y = 2x - a$, $\quad 2y + x = 3a$.
    At $(a, -a)$, $\quad y + 2x = a$, $\quad 2y = x - 3a$.

6. In the witch $y = \dfrac{8a^3}{4a^2 + x^2}$, find the equations of the tangent and normal at the point whose abscissa is $2a$.

    Ans. $x + 2y = 4a$, $\quad y = 2x - 3a$.

7. In the curve $\left(\dfrac{x}{a}\right)^{\frac{2}{3}} + \left(\dfrac{y}{b}\right)^{\frac{2}{3}} = 1$, find the equation of the tangent at the point (x', y'). Ans. $\dfrac{xx'}{a^2} + \dfrac{y + 2y'}{3b^{\frac{2}{3}} y'^{\frac{1}{3}}} = 1$.

8. In the ellipse $x^2 + 2y^2 - 2xy - x = 0$, find the equations of the tangent and normal at the points whose abscissa is 1.

    Ans. At $(1, 0)$, $\quad 2y = x - 1$, $\quad y + 2x = 2$.
    At $(1, 1)$, $\quad 2y = x + 1$, $\quad y + 2x = 3$.

9. In the parabola $x^{\frac{1}{2}} + y^{\frac{1}{2}} = a^{\frac{1}{2}}$, find the equation of the tangent at the point (x', y'). *Ans.* $xx'^{-\frac{1}{2}} + yy'^{-\frac{1}{2}} = a^{\frac{1}{2}}$.

10. Show that in the preceding curve the sum of the intercepts of the tangent on the co-ordinate axes is constant and equal to a.

11. In the hypocycloid $x^{\frac{2}{3}} + y^{\frac{2}{3}} = a^{\frac{2}{3}}$, find the equation of the tangent at the point (x', y'). *Ans.* $xx'^{-\frac{1}{3}} + yy'^{-\frac{1}{3}} = a^{\frac{2}{3}}$.

12. Show that in the preceding curve the part of the tangent intercepted between the co-ordinate axes is constant and equal to a.

100. *Asymptotes. Rectangular Co-ordinates.* When the tangent to a curve approaches a limiting position, as the distance of the point of contact from the origin is indefinitely increased, this limiting position is called an asymptote. In other words, an asymptote is a tangent which passes within a finite distance of the origin, although its point of contact is at an infinite distance.

101. From the equation of the tangent (1) Art. 99, we find for its intercepts on the co-ordinate axes,

$$\text{Intercept on } X = x' - y'\frac{dx'}{dy'},$$

$$\text{Intercept on } Y = y' - x'\frac{dy'}{dx'}.$$

If either of these intercepts is finite for $x' = \infty$, or $y' = \infty$, the corresponding tangent will be an asymptote.

The equation of this asymptote may be obtained from its two intercepts, or from one intercept and the limiting value of $\dfrac{dy'}{dx'}$.

ASYMPTOTES.

102. Omitting the accents in Art. 101 as no longer necessary, let us investigate the conic sections with reference to asymptotes.

(1). The parabola, $\quad y^2 = 4ax$.

Here $$\frac{dy}{dx} = \frac{2a}{y}.$$

Intercept on $X = x - y\frac{dx}{dy} = x - \frac{y^2}{2a} = -x$,

Intercept on $Y = y - x\frac{dy}{dx} = y - \frac{2ax}{y} = \frac{y}{2}$.

When $x = \infty$, $y = \infty$, and both intercepts are also infinite. Hence the parabola has no asymptote.

(2). The hyperbola, $\quad \frac{x^2}{a^2} - \frac{y^2}{b^2} = 1$.

Here $$\frac{dy}{dx} = \frac{b^2 x}{a^2 y}.$$

Intercept on $X = \frac{a^2}{x}$,

Intercept on $Y = -\frac{b^2}{y}$.

These intercepts are both zero when $x = \infty$, and there is an asymptote passing through the origin. To find its equation, it is necessary to find the value of $\frac{dy}{dx}$, when $x = \infty$.

$$\frac{dy}{dx} = \frac{b^2 x}{a^2 y} = \pm \frac{bx}{a\sqrt{x^2 - a^2}} = \pm \frac{b}{a} \frac{1}{\sqrt{1 - \frac{a^2}{x^2}}}.$$

Hence $\frac{dy}{dx} = \pm \frac{b}{a}$, when $x = \infty$.

There are then two asymptotes, whose equations are

$$y = \pm \frac{b}{a} x.$$

(3). The ellipse, having no infinite branches, can have no asymptote.

103. *Asymptotes Parallel to the Co-ordinate Axes.* When, in the equation of the curve, $x = \infty$ gives a finite value of y, as $y = a$, then $y = a$ is the equation of an asymptote parallel to X.

And when $y = \infty$ gives $x = a$, then $x = a$ is an asymptote parallel to Y.

104. *Asymptotes by Expansion.* Frequently an asymptote may be determined by solving the equation of the curve for x or y and expanding the second member.

For example, to find the asymptotes of the hyperbola

$$\frac{x^2}{a^2} - \frac{y^2}{b^2} = 1.$$

$$y = \pm \frac{b}{a}(x^2 - a^2)^{\frac{1}{2}} = \pm \frac{bx}{a}\left(1 - \frac{a^2}{x^2}\right)^{\frac{1}{2}} = \pm \frac{bx}{a}\left(1 - \frac{a^2}{2x^2} - \cdots\right).$$

As x increases indefinitely, the curve approaches the lines $y = \pm \dfrac{bx}{a}$, the asymptotes.

105. *Asymptotes. Polar Co-ordinates.* From the figure of Art. 97, it is evident that for an asymptote, the polar subtangent OT has a finite limit, as OP is indefinitely increased. That is, when $r^2 \dfrac{d\theta}{dr}$ has a finite limit for $r = \infty$, there is an asymptote at that distance from the pole, and parallel to r.

If the distance $r^2 \dfrac{d\theta}{dr}$ is positive, it is to the right, and if negative, to the left, of the pole, looking in the direction of the infinite r.

ASYMPTOTES.

106. For example, find the asymptotes of the curve

$$r = a \tan \theta.$$

Here $\dfrac{dr}{d\theta} = a \sec^2 \theta,$

and the subtangent $= r^2 \dfrac{d\theta}{dr}$

$= a \sin^2 \theta.$

When $\theta = \pm \dfrac{\pi}{2},$

we have $r = \infty,$

and the subtangent $= a.$

There are two asymptotes perpendicular to OX, at the distance a from the pole, on each side of it.

EXAMPLES.

Investigate the following curves with reference to asymptotes:

1. $y = \dfrac{x^3}{x^2 + 3a^2}.$ Asymptote, $y = x.$

2. $y^3 = 6x^2 - x^3.$ Asymptote, $x + y = 2.$

3. The cissoid $y^2 = \dfrac{x^3}{2a - x}.$ Asymptote, $x = 2a.$

4. $x^3 + y^3 = a^3.$ Asymptote, $x + y = 0.$

5. $(x - 2a)y^2 = x^3 - a^3.$ Asymptotes, $x = 2a,\ x + a = \pm y.$

6. $-x^3 + y^3 = 3axy.$ Asymptote, $x + y + a = 0.$
 (Substitute $y = vx$ in the given equation and in the expressions for the intercepts.)

7. The reciprocal spiral $r = \dfrac{a}{\theta}.$
 Asymptote parallel to OX, at the distance a above.

8. $r = a \sec 2\theta$.

There are four asymptotes at the same distance $\frac{a}{2}$ from the pole, and inclined 45° with OX.

9. The parabola $r = \dfrac{a}{1 - \cos \theta}$.   There is no asymptote.

10. $(r - a) \sin \theta = b$.

There is an asymptote parallel to OX, at the distance b above.

11. $r = a(\sec 2\theta + \tan 2\theta)$.

There are two asymptotes parallel to $\theta = \dfrac{\pi}{4}$, at the distance a on each side of the pole.

CHAPTER XIII.

DIRECTION OF CURVATURE. POINTS OF INFLEXION.

107. A curve is either concave upward or concave downward. It will now be shown that when the equation of the curve is in rectangular co-ordinates, the curve is concave *upward* or *downward*, according as $\frac{d^2y}{dx^2}$ is *positive* or *negative*.

108. *Lemma.* If u is a function of x which *increases* as x increases, then $\frac{du}{dx} > 0$; but if u *decreases* as x increases, $\frac{du}{dx} < 0$.

For, in the former case Δu and Δx have the *same* sign, and therefore $\frac{\Delta u}{\Delta x} > 0$, and consequently $\frac{du}{dx} > 0$.

In the latter case, Δu and Δx have *different* signs, and therefore $\frac{\Delta u}{\Delta x} < 0$, and $\frac{du}{dx} < 0$.

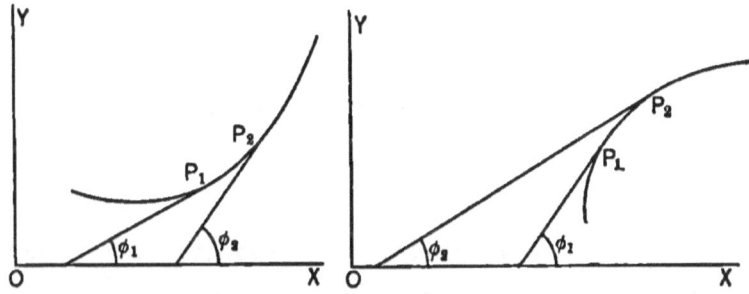

109. By inspection of the first of the two figures above, we see that when the curve is concave *upward*, ϕ *increases* as x increases, and consequently $\tan \phi$ *increases* as x increases.

Hence, by Art. 108, $\quad \dfrac{d \tan \phi}{dx} > 0$;

that is, $\qquad \dfrac{d}{dx}\left(\dfrac{dy}{dx}\right) > 0 \text{ or } \dfrac{d^2y}{dx^2} > 0.$

From the second figure, we see that when the curve is concave *downward*, tan ϕ *decreases* as x increases, and therefore

$$\frac{d \tan \phi}{dx} < 0;$$

that is,
$$\frac{d^2y}{dx^2} < 0.$$

110. *A Point of Inflexion* of a curve is a point P, where the curvature changes, the curve on one side of this point being concave upward, and on the other, concave downward. Hence, by Art. 109, at a point of inflexion, $\frac{d^2y}{dx^2}$ changes sign; that is,

$$\frac{d^2y}{dx^2} = 0 \text{ or } \infty.$$

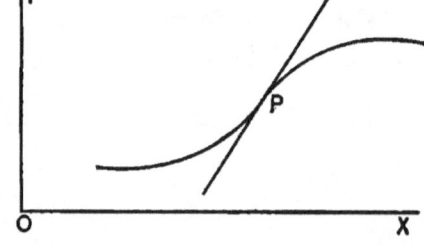

It is evident that the tangent at a point of inflexion intersects the curve at that point.

Find the point of inflexion of the curve $y = (x-1)^3$, and the direction of curvature on each side of it.

Here
$$\frac{d^2y}{dx^2} = 6(x-1).$$

Putting this equal to zero, we have for the required point of inflexion, $x = 1$. If $x < 1$, $\frac{d^2y}{dx^2} < 0$; and if $x > 1$, $\frac{d^2y}{dx^2} > 0$.

Hence the curve is concave downward on the left, and concave upward on the right, of the point of inflexion.

EXAMPLES.

Find the points of inflexion, and the direction of curvature, of the three following curves: —

1. The curve $a^2y = \frac{x^3}{3} - ax^2 + 2a^3$.

   Ans. $\left(a, \frac{4a}{3}\right)$; concave downward on the left of this point, concave upward on the right.

2. The witch $y = \dfrac{8a^3}{x^2 + 4a^2}$.

Ans. $\left(\pm \dfrac{2a}{\sqrt{3}}, \dfrac{3a}{2}\right)$; concave downward between these points, concave upward outside of them.

3. The curve $y = \dfrac{x^3}{x^2 + 3a^2}$.

Ans. $\left(-3a, -\dfrac{9a}{4}\right)$, $(0, 0)$, $\left(3a, \dfrac{9a}{4}\right)$; concave upward on the left of first point, downward between first and second, upward between second and third, and downward on the right of third point.

4. Find the points of inflexion of the curve $\left(\dfrac{x}{a}\right)^2 + \left(\dfrac{y}{b}\right)^{\frac{2}{3}} = 1$.

Ans. $x = \pm \dfrac{a}{\sqrt{2}}$.

5. Find the points of inflexion of the curve $a^4 y^2 = a^2 x^4 - x^6$.

Ans. $x = \pm \dfrac{a}{6} \sqrt{27 - 3\sqrt{33}}$.

CHAPTER XIV.

CURVATURE. RADIUS OF CURVATURE. EVOLUTE AND INVOLUTE.

111. *Definition of Curvature.* If a point moves in a straight line, the direction of its motion is the same at every point of its course; but if its path is a curved line, there is a continual change of direction as it moves along the curve. This change of direction is called *curvature*.

The direction at any point being the same as that of the tangent at that point, the curvature may be determined by comparing the linear motion of the point with the simultaneous angular motion of the tangent. The curvature is either uniform or variable.

112. *Uniform Curvature.* The curvature is uniform when, as the point moves over equal arcs, the tangent turns through equal angles. It is then measured by the angle described by the tangent while the point describes a unit of arc.

Suppose the point P to move in the curve AQ. Let $s = AP$ denote its distance along the curve from any fixed point A, and let $\phi = PTX$, the angle made by the tangent PT with the fixed line OX. Then as the point describes the arc PQ, which is denoted by Δs, the tangent turns through the angle QRK or $\Delta\phi$. Then, if the curvature is uniform, it is equal to $\dfrac{\Delta\phi}{\Delta s}$.

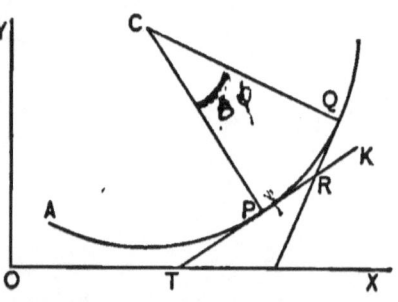

The circle is the only curve of uniform curvature. Supposing APQ an arc of a circle, if we draw the radii CP and CQ, and let r denote the length of the radius, then the angle $PCQ = QRK = \Delta\phi$; but arc $PQ = CP \times$ angle PCQ; that is, $\Delta s = r\Delta\phi$.

RADIUS OF CURVATURE. 121

Hence $r = \dfrac{\Delta s}{\Delta \phi}$; that is, *the radius of a circle is the reciprocal of its curvature.*

113. Variable Curvature. In this case the tangent does *not* turn through equal angles as the point describes equal arcs. Here $\dfrac{\Delta \phi}{\Delta s}$ is the mean curvature throughout the arc Δs. The curvature at the beginning of this arc is more nearly equal to $\dfrac{\Delta \phi}{\Delta s}$, the shorter we take Δs. Hence the curvature at any point is the limit of $\dfrac{\Delta \phi}{\Delta s}$, that is, $\dfrac{d\phi}{ds}$.

114. Radius of Curvature. A circle tangent to a curve at any point, and having the same curvature as that of the curve at that point, is called the *circle of curvature;* its radius, the *radius of curvature;* and its centre, the *centre of curvature.*

The curvature of this circle being that of the given curve, is equal to $\dfrac{d\phi}{ds}$. If we denote the radius of curvature by ρ, then by Art. 112,

$$\rho = \frac{ds}{d\phi}. \quad \cdots \cdots \cdots \quad (1)$$

To obtain ρ in terms of x and y, we have from (1), Art. 98,

$$\frac{ds}{dx} = \sqrt{1 + \left(\frac{dy}{dx}\right)^2}.$$

From (1), Art. 94, $\tan \phi = \dfrac{dy}{dx}$, $\phi = \tan^{-1}\left(\dfrac{dy}{dx}\right)$.

Differentiating, $\quad \dfrac{d\phi}{dx} = \dfrac{\dfrac{d^2y}{dx^2}}{1 + \left(\dfrac{dy}{dx}\right)^2}. \quad \cdots \cdots \quad (2)$

Hence $\quad \rho = \dfrac{ds}{d\phi} = \dfrac{\dfrac{ds}{dx}}{\dfrac{d\phi}{dx}} = \dfrac{\left[1 + \left(\dfrac{dy}{dx}\right)^2\right]^{\frac{3}{2}}}{\dfrac{d^2y}{dx^2}}. \quad \cdots \quad (3)$

Also, by interchanging x and y, we have

$$\rho = \frac{\left[1+\left(\dfrac{dx}{dy}\right)^2\right]^{\frac{3}{2}}}{\dfrac{d^2x}{dy^2}},$$

which is sometimes the more convenient expression.

As an example, find the radius of curvature of the semi-cubical parabola $ay^2 = x^3$.

Differentiating, $\dfrac{dy}{dx} = \dfrac{3x^{\frac{1}{2}}}{2a^{\frac{1}{2}}}$, $\dfrac{d^2y}{dx^2} = \dfrac{3}{4(ax)^{\frac{1}{2}}}$.

Substituting in (3), we find

$$\rho = \frac{x^{\frac{1}{2}}(4a+9x)^{\frac{3}{2}}}{6a}.$$

EXAMPLES.

Find the radius of curvature of the following curves:—

1. The parabola $y^2 = 4ax$.   Ans. $\rho = \dfrac{2(x+a)^{\frac{3}{2}}}{a^{\frac{1}{2}}} = \dfrac{2a}{\sin^3\phi}$.

2. The equilateral hyperbola $2xy = a^2$.   Ans. $\rho = \dfrac{(x^2+y^2)^{\frac{3}{2}}}{a^2}$.

3. The ellipse $\dfrac{x^2}{a^2} + \dfrac{y^2}{b^2} = 1$.   Ans. $\rho = \dfrac{(a^4y^2 + b^4x^2)^{\frac{3}{2}}}{a^4b^4}$.

What are the values of ρ at the extremities of the major and minor axes?   Ans. $\dfrac{b^2}{a}$ and $\dfrac{a^2}{b}$.

4. The curve $\left(\dfrac{x}{a}\right)^{\frac{2}{3}} + \left(\dfrac{y}{b}\right)^{\frac{2}{3}} = 1$, at the point $(0, b)$.

Ans. $\rho = \dfrac{a^2}{3b}$.

5. The curve $y = \log \sec x$.   Ans. $\rho = \sec x$.

6. The parabola $x^{\frac{1}{2}} + y^{\frac{1}{2}} = a^{\frac{1}{2}}$.   Ans. $\rho = \dfrac{2(x+y)^{\frac{3}{2}}}{a^{\frac{1}{2}}}$.

7. The catenary $y = \dfrac{a}{2}(e^{\frac{x}{a}} + e^{-\frac{x}{a}})$.   *Ans.* $\rho = \dfrac{y^2}{a}$.

8. The hypocycloid $x^{\frac{2}{3}} + y^{\frac{2}{3}} = a^{\frac{2}{3}}$.   *Ans.* $\rho = 3(axy)^{\frac{1}{3}}$.

9. The curve $a^4 y^2 = a^2 x^4 - x^6$, at the points $(0, 0)$ and $(a, 0)$.

$$\text{\textit{Ans.} } \rho = \frac{a}{2} \text{ and } \rho = a.$$

10. The cissoid $y^2 = \dfrac{x^3}{2a - x}$.   *Ans.* $\rho = \dfrac{ax^{\frac{1}{2}}(8a - 3x)^{\frac{3}{2}}}{3(2a - x)^2}$.

115. *Radius of Curvature in Polar Co-ordinates.* Resuming (1) Art. 114, $\rho = \dfrac{ds}{d\phi}$, let us express ρ in terms of r and θ.

From (3) Art. 98½, $\dfrac{ds}{d\theta} = \sqrt{r^2 + \left(\dfrac{dr}{d\theta}\right)^2}$.

From (c) Art. 96,

$$\phi = \theta + \psi, \quad \therefore \quad \frac{d\phi}{d\theta} = 1 + \frac{d\psi}{d\theta}.$$

From (b) Art. 96,

$$\tan \psi = \frac{r}{\dfrac{dr}{d\theta}}, \text{ or } \psi = \tan^{-1}\!\left(\frac{r}{\dfrac{dr}{d\theta}}\right).$$

Differentiating, $\dfrac{d\psi}{d\theta} = \dfrac{\left(\dfrac{dr}{d\theta}\right)^2 - r\dfrac{d^2r}{d\theta^2}}{r^2 + \left(\dfrac{dr}{d\theta}\right)^2}$.

Substituting, $\dfrac{d\phi}{d\theta} = \dfrac{r^2 + 2\left(\dfrac{dr}{d\theta}\right)^2 - r\dfrac{d^2r}{d\theta^2}}{r^2 + \left(\dfrac{dr}{d\theta}\right)^2}$.

Hence $\rho = \dfrac{\dfrac{ds}{d\theta}}{\dfrac{d\phi}{d\theta}} = \dfrac{\left[r^2 + \left(\dfrac{dr}{d\theta}\right)^2\right]^{\frac{3}{2}}}{r^2 + 2\left(\dfrac{dr}{d\theta}\right)^2 - r\dfrac{d^2r}{d\theta^2}}$ . . . (1)

EXAMPLES.

Find the radius of curvature of the following curves: —

1. The circle $r = a \sin \theta$.   Ans. $\rho = \dfrac{a}{2}$.

2. The logarithmic spiral $r = e^{a\theta}$.   Ans. $\rho = r\sqrt{1+a^2}$.

3. The spiral of Archimedes $r = a\theta$.   Ans. $\rho = \dfrac{(r^2+a^2)^{\frac{3}{2}}}{r^2+2a^2}$.

4. The cardioid $r = a(1-\cos\theta)$.   Ans. $\rho^2 = \dfrac{8}{9}ar$.

5. The curve $r = a \sin^3 \dfrac{\theta}{3}$.   Ans. $\rho = \dfrac{3}{4}a \sin^2 \dfrac{\theta}{3}$.

6. The parabola $r = a \sec^2 \dfrac{\theta}{2}$.   Ans. $\rho = 2a \sec^3 \dfrac{\theta}{2}$.

7. The lemniscate $r^2 = a^2 \cos 2\theta$.   Ans. $\rho = \dfrac{a^2}{3r}$.

116. *Co-ordinates of the Centre of Curvature.* Let x, y be the co-ordinates of P, any point of the curve AB, and C the corresponding centre of curvature. CP is then the radius of curvature, and is normal to the curve.

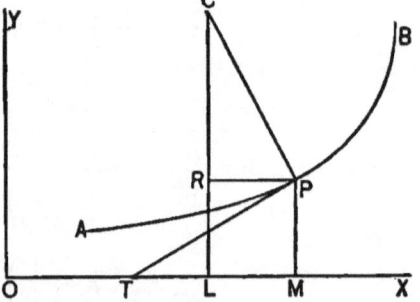

Draw also the tangent PT.

Then $CP = \rho$;

angle $PCR = PTX = \phi$.

Let α, β, be the co-ordinates of C.

$$OL = OM - RP, \quad LC = MP + RC;$$

that is, $\alpha = x - \rho \sin \phi, \quad \beta = y + \rho \cos \phi$ . . . (1)

To express α and β in terms of x and y, we have, by (2) Art. 98, and (1), (2), Art. 114,

$$\rho \sin \phi = \frac{ds}{d\phi}\frac{dy}{ds} = \frac{dy}{d\phi} = \frac{dy}{dx}\frac{dx}{d\phi} = \frac{\frac{dy}{dx}\left[1+\left(\frac{dy}{dx}\right)^2\right]}{\frac{d^2y}{dx^2}},$$

$$\rho \cos \phi = \frac{ds}{d\phi}\frac{dx}{ds} = \frac{dx}{d\phi} = \frac{1+\left(\frac{dy}{dx}\right)^2}{\frac{d^2y}{dx^2}}.$$

Hence
$$a = x - \frac{\frac{dy}{dx}\left[1+\left(\frac{dy}{dx}\right)^2\right]}{\frac{d^2y}{dx^2}}, \quad \beta = y + \frac{1+\left(\frac{dy}{dx}\right)^2}{\frac{d^2y}{dx^2}} \quad \cdots \quad (2)$$

117. *Evolute and Involute.* Every point of a curve AB has a corresponding centre of curvature. Thus, P_1, P_2, P_3, etc., have for their respective centres of curvature C_1, C_2, C_3, etc. The curve HK, which is the locus of the centres of curvature, is called the *evolute* of AB.

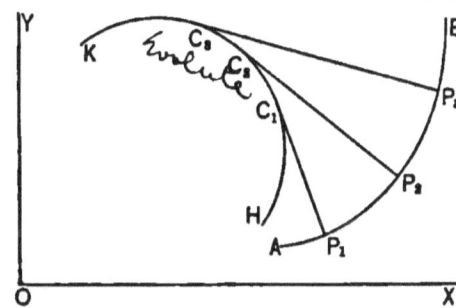

To express the inverse relation, AB is called the *involute* of HK.

118. *To find the equation of the evolute of a given curve.*

By (2) Art. 116, a and β, the co-ordinates of any point of the required evolute, may be expressed in terms of x and y, the co-ordinates of any point of the given curve. These two equations, together with that of the given curve, furnish three equations between a, β, x, and y, from which, if x and y are eliminated, we obtain a relation between a and β, which is the equation of the required evolute.

For example, find the equation of the evolute of the parabola $y^2 = 4ax$.

Here $\quad \dfrac{dy}{dx} = a^{\frac{1}{2}} x^{-\frac{1}{2}}, \quad \dfrac{d^2y}{dx^2} = -\dfrac{1}{2} a^{\frac{1}{2}} x^{-\frac{3}{2}}.$

Substituting in (2) Art. 116, we have

$$a = 3x + 2a, \quad \beta = -\frac{2x^{\frac{3}{2}}}{a^{\frac{1}{2}}}.$$

Eliminating x, we have for the equation of the evolute,

$$a\beta^2 = \frac{4}{27}(a - 2a)^3.$$

This curve is the semi-cubical parabola. The figure shows its form and position. F is the focus of the given parabola.

$$OC = 2a = 2 \times OF.$$

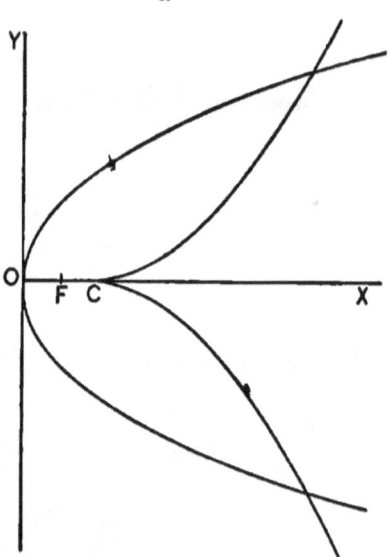

119. *Properties of the Involute and Evolute.* Let us return to the equations, (1) Art. 116,

$$a = x - \rho \sin\phi,$$
$$\beta = y + \rho \cos\phi.$$

Differentiating with reference to s, and by (2) Art. 98, and (1) Art. 114, we have

$$\frac{da}{ds} = \frac{dx}{ds} - \frac{d\rho}{ds}\sin\phi - \rho\cos\phi\frac{d\phi}{ds} = -\frac{d\rho}{ds}\sin\phi \quad \ldots \quad (a)$$

$$\frac{d\beta}{ds} = \frac{dy}{ds} + \frac{d\rho}{ds}\cos\phi - \rho\sin\phi\frac{d\phi}{ds} = \frac{d\rho}{ds}\cos\phi \quad \ldots \quad (b)$$

Dividing (b) by (a),

$$\frac{d\beta}{da} = -\cot\phi = \tan\left(\phi + \frac{\pi}{2}\right).$$

If ϕ' denote the angle made with the axis of X by the tangent to the evolute, then, by (1) Art. 94,

$$\frac{d\beta}{da} = \tan\phi'. \quad \therefore \quad \phi' = \phi + \frac{\pi}{2}.$$

That is, the tangent to the evolute is perpendicular to the corresponding tangent to the involute. In other words, a tangent to the evolute at any point C_1 (Fig. Art. 117), is C_1P_1, the normal to the involute at P_1.

120. Again, from (a) and (b), Art. 119,

$$\left(\frac{d\alpha}{ds}\right)^2 + \left(\frac{d\beta}{ds}\right)^2 = \left(\frac{d\rho}{ds}\right)^2, \text{ or } \left(\frac{ds'}{ds}\right)^2 = \left(\frac{d\rho}{ds}\right)^2,$$

where s' denotes the length of the arc of the evolute measured from a fixed point. Hence,

$$\frac{ds'}{ds} = \pm \frac{d\rho}{ds}, \text{ and therefore } \Delta s' = \pm \Delta \rho. \quad \text{\textit{; why not}}$$

That is, the difference between any two radii of curvature P_1C_1, P_3C_3, is equal to the corresponding included arc of the evolute C_1C_3.

121. From the two properties of Arts. 119 and 120, it follows that the involute AB may be described by the end of a string unwound from the evolute HK. From this property the word *evolute* is derived.

It will be noticed that a curve has only one evolute, but an infinite number of involutes, as may be seen by varying the length of the string which is unwound. Such curves are called *parallel* curves.

EXAMPLES.

1. Find the co-ordinates of the centre of curvature of the cubical parabola $y^3 = a^2x$.

$$Ans. \ \alpha = \frac{a^4 + 15\,y^4}{6\,a^2y}, \quad \beta = \frac{a^4y - 9\,y^5}{2\,a^4}.$$

2. Find the co-ordinates of the centre of curvature of the catenary $y = \frac{a}{2}(e^{\frac{x}{a}} + e^{-\frac{x}{a}})$.

$$Ans. \ \alpha = x - \frac{y}{a}\sqrt{y^2 - a^2}, \quad \beta = 2y.$$

3. Find the co-ordinates of the centre of curvature, and the equation of the evolute, of the ellipse $\frac{x^2}{a^2} + \frac{y^2}{b^2} = 1$.

$$Ans. \ \alpha = \frac{(a^2 - b^2)x^3}{a^4}, \quad \beta = -\frac{(a^2 - b^2)y^3}{b^4}$$

$$(a\alpha)^{\frac{2}{3}} + (b\beta)^{\frac{2}{3}} = (a^2 - b^2)^{\frac{2}{3}}.$$

4. Show that in the parabola $x^{\frac{1}{2}} + y^{\frac{1}{2}} = a^{\frac{1}{2}}$ we have the relation $\alpha + \beta = 3(x + y)$.

5. Find the co-ordinates of the centre of curvature, and the equation of the evolute, of the hypocycloid $x^{\frac{2}{3}} + y^{\frac{2}{3}} = a^{\frac{2}{3}}$.

$$\text{Ans. } \alpha = x + 3x^{\frac{1}{3}}y^{\frac{2}{3}}, \quad \beta = y + 3x^{\frac{2}{3}}y^{\frac{1}{3}}.$$
$$(\alpha + \beta)^{\frac{2}{3}} + (\alpha - \beta)^{\frac{2}{3}} = 2a^{\frac{2}{3}}.$$

6. Given the equation of the equilateral hyperbola $2xy = a^2$; show that
$$\alpha + \beta = \frac{(y + x)^3}{a^2}, \quad \alpha - \beta = \frac{(y - x)^3}{a^2}.$$
Thence derive the equation of the evolute
$$(\alpha + \beta)^{\frac{2}{3}} - (\alpha - \beta)^{\frac{2}{3}} = 2a^{\frac{2}{3}}.$$

CHAPTER XV.

ORDER OF CONTACT. OSCULATING CIRCLE.

122. *Definition.* Suppose two curves to have two common points P_1 and P_2. If one of these points, as P_2, be supposed to approach to coincidence with P_1, the limiting position is called a contact of the *first order*. Thus two curves are said to have contact of the *first order* when they have *two* consecutive common points.

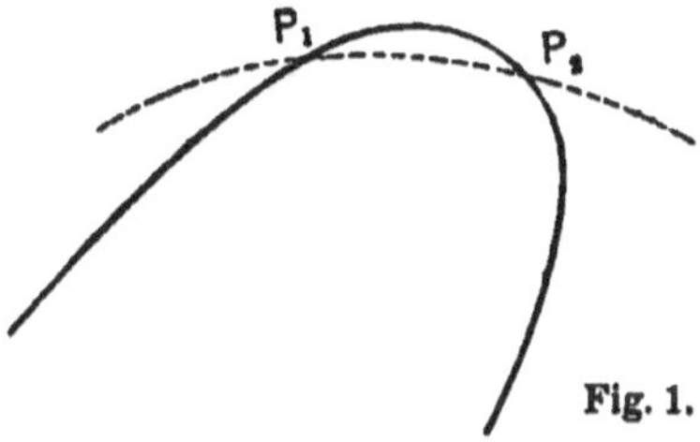

Again, suppose the two curves, having at P a contact of the first order, to have a third common point P_3. Now when P_3 moves up to coincidence with P, we have ultimately a contact of the *second order*, which thus denotes *three* consecutive common points.

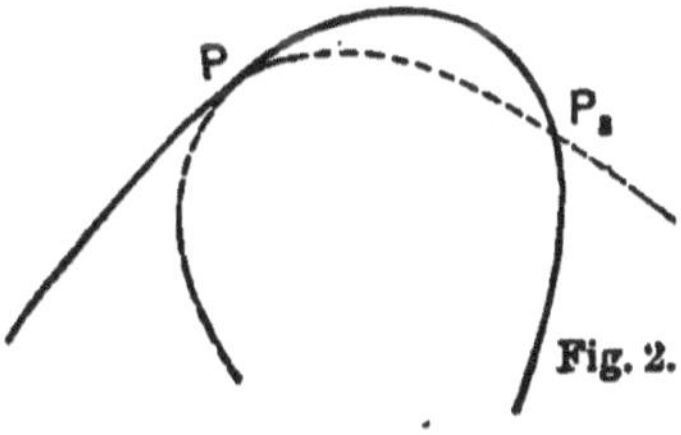

Similarly, suppose the two curves to have three common consecutive points at P, forming a contact of the second order, and a fourth common point P_4. By supposing P_4 to move up to P, we have a contact of the *third order*, containing *four* consecutive common points.

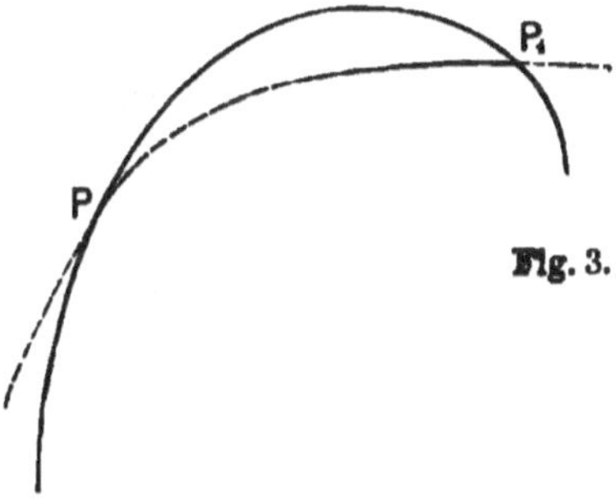

In general, a contact of the nth *order* includes $n+1$ consecutive common points.

Fig. 1.

Fig. 2.

Fig. 3.

123. *When the order of contact is even, the curves cross at the point of contact; but when the order is odd, they do not cross.*

For a contact of the first order, it is evident from Fig. 1, Art. 122, that outside of P_1 and P_2, the dotted curve is on the same side of the other curve. Hence, when the two points coincide to form the point of contact, the curves do not cross at that point.

For a contact of the second order, it is evident from Fig. 2, Art. 122, that when P_3 coincides with P, the curves cross at the point of contact.

For a contact of the third order, Fig. 3, Art. 122 shows that the curves do not cross at the point of contact.

Similarly it is evident that the proposition is generally true.

124. *Osculating Curves.* As a straight line can be made to pass through only two points, the *tangent* has generally a contact of only the first order with a curve.

The circle having the closest contact with a curve at a given point is called the *osculating circle*. As a circle can be made to pass through only three points, the osculating circle has generally contact of the second order.

The parabola of closest contact is likewise called the *osculating parabola*. As a parabola can be made to pass through four points, the osculating parabola has contact of the third order.

The conic of closest contact is called the *osculating conic*.

As a conic can be made to pass through five points, the osculating conic has contact of the fourth order.

It is evident from Art. 123 that the osculating circle and osculating conic cross the curve at the point of contact, while the tangent and osculating parabola do not.

125. *Exceptional Points.* Although the tangent has generally contact of the first order, it may at exceptional points of a curve have a contact of a higher order.

For example, since the tangent at a point of inflexion crosses the curve, it follows from Art. 123, that the order of contact must be even. Hence at a point of inflexion the tangent has contact of at least the second order.

The osculating circle, which has generally contact of the second order, has a higher order of contact at points of maximum or minimum curvature, as, for example, the vertices of an ellipse. It is evident from the symmetry of the ellipse with reference to its vertices, that no circle tangent at these points would cross the curve at the point of contact. Hence, by Art. 123, the order of contact is odd, — at least the third.

126. *Analytical Conditions for Contact.*

Let $\qquad y = \phi(x), \quad \text{and} \quad y = \psi(x),$

be the equations of two curves having two common points P and Q.

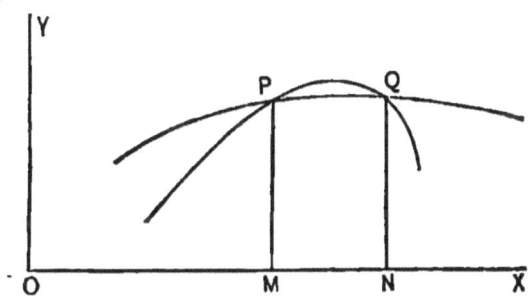

Let $\qquad OM = a, \quad MN = h.$

Then $\qquad \phi(a) = \psi(a), \quad \text{and} \quad \phi(a+h) = \psi(a+h).$

Expanding each member of this equation by Taylor's Theorem,

$$\phi(a) + h\phi'(a) + \frac{h^2}{2}\phi''(a) + \frac{h^3}{\underline{|3}}\phi'''(a) + \cdots$$
$$= \psi(a) + h\psi'(a) + \frac{h^2}{2}\psi''(a) + \frac{h^3}{\underline{|3}}\psi'''(a) + \cdots. \quad \cdot \quad \cdot \quad \cdot \quad (1)$$

Since $\phi(a)=\psi(a)$, we have from (1) after dividing by h,

$$\phi'(a)+\frac{h}{2}\phi''(a)+\cdots=\psi'(a)+\frac{h}{2}\psi''(a)+\cdots.$$

When Q approaches P, h approaches zero, and we have at the limit
$$\phi'(a)=\psi'(a).$$

Hence the conditions for a contact of the first order at the point $x=a$, are
$$\phi(a)=\psi(a), \quad \phi'(a)=\psi'(a).$$

127. Again, suppose the two curves have a contact of the first order at P and another common point Q.

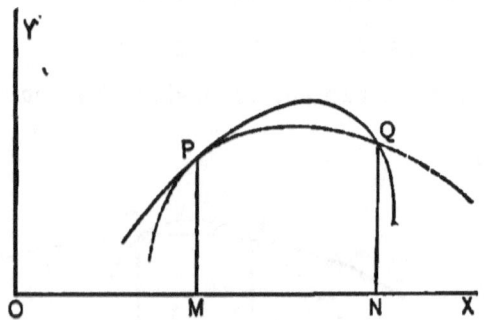

As before, let $OM=a$, $MN=h$.

Since $\phi(a)=\psi(a)$, and $\phi'(a)=\psi'(a)$,

we have from (1) Art. 126, after dividing by h^2,

$$\frac{1}{2}\phi''(a)+\frac{h}{\underline{|3}}\phi'''(a)+\cdots=\frac{1}{2}\psi''(a)+\frac{h}{\underline{|3}}\psi'''(a)+\cdots.$$

When Q approaches P, we have at the limit, when $h=0$,
$$\phi''(a)=\psi''(a).$$

Hence the conditions for a contact of the second order at the point $x=a$, are
$$\phi(a)=\psi(a), \quad \phi'(a)=\psi'(a), \quad \phi''(a)=\psi''(a).$$

OSCULATING CIRCLE. 133

128. *Conditions for contact of the nth order.* The same process may be extended to contacts of higher orders, every additional point in the contact adding one to the series of equalities at the end of the preceding article.

In general, the conditions for a contact of the nth order at the point $x = a$, are

$$\phi(a)=\psi(a), \quad \phi'(a)=\psi'(a), \quad \phi''(a)=\psi''(a), \quad \cdots \quad \phi^n(a)=\psi^n(a).$$

In other words, for $x = a$,

$$y, \quad \frac{dy}{dx}, \quad \frac{d^2y}{dx^2}, \quad \cdots \quad \frac{d^ny}{dx^n},$$

must all have the same values, respectively, taken from the equations of both curves.

129. *To find the co-ordinates of the centre, and radius, of the osculating circle at any point of a given curve.*

Let the equation of the given curve be

$$y = f(x).$$

The general equation of a circle with centre (a, b) and radius r, is

$$(x-a)^2 + (y-b)^2 = r^2. \quad \cdots \cdots \quad (1)$$

Differentiating twice successively, we have

$$x - a + (y-b)\frac{dy}{dx} = 0, \quad \cdots \cdots \quad (2)$$

$$1 + \left(\frac{dy}{dx}\right)^2 + (y-b)\frac{d^2y}{dx^2} = 0. \quad \cdots \cdots \quad (3)$$

From (3), $\quad y - b = -\dfrac{1+\left(\dfrac{dy}{dx}\right)^2}{\dfrac{d^2y}{dx^2}}. \quad \cdots \cdots \quad (4)$

From (2), $\quad x - a = \dfrac{\dfrac{dy}{dx}\left[1+\left(\dfrac{dy}{dx}\right)^2\right]}{\dfrac{d^2y}{dx^2}}. \quad \cdots \cdots \quad (5)$

Substituting (4) and (5) in (1),

$$r^2 = \frac{\left[1+\left(\frac{dy}{dx}\right)^2\right]^3}{\left(\frac{d^2y}{dx^2}\right)^2}. \quad \ldots \ldots \ldots (6)$$

Hence
$$a = x - \frac{\frac{dy}{dx}\left[1+\left(\frac{dy}{dx}\right)^2\right]}{\frac{d^2y}{dx^2}}, \quad b = y + \frac{1+\left(\frac{dy}{dx}\right)^2}{\frac{d^2y}{dx^2}}, \quad (7)$$

and
$$r = \frac{\left[1+\left(\frac{dy}{dx}\right)^2\right]^{\frac{3}{2}}}{\frac{d^2y}{dx^2}}. \quad \ldots \ldots \ldots (8)$$

In these expressions, x, y, $\frac{dy}{dx}$, $\frac{d^2y}{dx^2}$, refer to (1), the equation of the circle; but since the osculating circle by definition has contact of the second order with the given curve, these quantities will have the same values if derived from the equation $y = f(x)$, at the point of contact.

By comparing (7) and (8) with the expressions for a, β, and ρ, in Arts. 114, 116, it is evident that the osculating circle is the same as the circle of curvature.

130. *At a point of maximum or minimum curvature, the osculating circle has contact of the third order.*

If we regard equation (8) in the preceding article as referring to the given curve, $y = f(x)$, we have as a condition for a maximum or minimum value of r,

$$\frac{dr}{dx} = 0. \qquad \text{(See Art. 146.)}$$

We thus obtain from (8),

$$3\frac{dy}{dx}\left(\frac{d^2y}{dx^2}\right)^2 - \left[1+\left(\frac{dy}{dx}\right)^2\right]\frac{d^3y}{dx^3} = 0,$$

from which
$$\frac{d^3y}{dx^3} = \frac{3\frac{dy}{dx}\left(\frac{d^2y}{dx^2}\right)^2}{1+\left(\frac{dy}{dx}\right)^2}. \quad \ldots \ldots (1)$$

Again, if we regard (8) as referring to the osculating circle
$$(x-a)^2 + (y-b)^2 = r^2,$$
we shall also have
$$\frac{dr}{dx} = 0,$$
since r is constant for all points on the circle.

Thus we obtain, both for the curve and the circle, the same expression (1) for $\frac{d^3y}{dx^3}$, and since $\frac{dy}{dx}$ and $\frac{d^2y}{dx^2}$ in the second member of (1) have, at the point of contact, the same values for both curves, it follows that $\frac{d^3y}{dx^3}$ has likewise the same value. Hence the contact is of the third order.

EXAMPLES.

1. Find the order of contact of the two curves,
$$y = x^3, \quad \text{and} \quad y = 3x^2 - 3x + 1.$$

By combining the two equations, the point, $x=1$, $y=1$, is found to be common to both curves.

Differentiating the two given equations,

$$y = x^3, \qquad\qquad y = 3x^2 - 3x + 1,$$

$$\frac{dy}{dx} = 3x^2, \qquad\qquad \frac{dy}{dx} = 6x - 3,$$

$$\frac{d^2y}{dx^2} = 6x, \qquad\qquad \frac{d^2y}{dx^2} = 6,$$

$$\frac{d^3y}{dx^3} = 6, \qquad\qquad \frac{d^3y}{dx^3} = 0.$$

When $x = 1$, $\dfrac{dy}{dx} = 3$, in both curves;

when $x = 1$, $\dfrac{d^2y}{dx^2} = 6$, in both curves;

but $\dfrac{d^3y}{dx^3}$ has different values in the two curves.

Hence the contact is of the second order.

2. Find the order of contact of the parabola $4y = x^2$, and the straight line $y = x - 1$.   *Ans.* First order.

3. Find the order of contact of
$$9y = x^3 - 3x^2 + 27, \quad \text{and} \quad 9y + 3x = 28.$$
Ans. Second order.

4. Find the order of contact of
$$y = \log(x - 1), \quad \text{and} \quad x^2 - 6x + 2y + 8 = 0,$$
at the common point $(2, 0)$.   *Ans.* Second order.

5. Find the order of contact of the parabola $4y = x^2 - 4$, and the circle $x^2 + y^2 - 2y = 3$.   *Ans.* Third order.

6. What must be the value of a, in order that the parabola
$$y = x + 1 + a(x - 1)^2,$$
may have contact of the second order with the hyperbola
$$xy = 3x - 1 ?$$
Ans. $a = -1$.

7. Find the order of contact of the parabola
$$(x - 2a)^2 + (y - 2a)^2 = 2xy,$$
and the hyperbola $xy = a^2$.   *Ans.* Third order.

CHAPTER XVI.

ENVELOPES.

131. *Series of Curves.* When, in the equation of a curve, different values are assigned to one of its constants, the resulting equations represent a series of curves, differing in position, but all of the same kind or family.

For example, if we give different values to a in the equation of the parabola $y^2 = 4ax$, we obtain a series of parabolas, all having a common vertex and axis, but different focal distances.

Again, take the equation of the circle $(x-a)^2 + (y-b)^2 = c^2$. By giving different values to a, we have a series of equal circles whose centres are on the line $y = b$.

The quantity a which remains constant for any one curve of the series, but varies as we pass from one curve to another, is called the *parameter* of the series.

Sometimes two parameters are supposed to vary simultaneously, so as to satisfy a given relation between them.

Thus, in the equation of the circle $(x-a)^2 + (y-b)^2 = c^2$, we may suppose a and b to vary, subject to the condition,

$$a^2 + b^2 = k^2.$$

We then have a series of equal circles, whose centres are on another circle described about the origin with radius k.

132. *Definition of Envelope.* The intersection of any two curves of a series will approach a certain limit, as the two curves approach coincidence. Now, if we suppose the parameter to vary by infinitesimal increments, the locus of the ultimate intersections of consecutive curves is called the *envelope* of the series.

133. *The envelope of a series of curves is tangent to every curve of the series.*

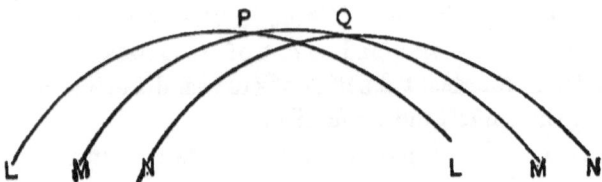

Suppose L, M, N to be any three curves of the series. P is the intersection of M with the preceding curve L, and Q its intersection with the following curve N.

As the curves approach coincidence, P and Q will ultimately be two consecutive points of the envelope, and of the curve M. Hence the envelope touches M.

Similarly, it may be shown that the envelope touches any other curve of the series.

134. *To find the equation of the envelope of a given series of curves.*

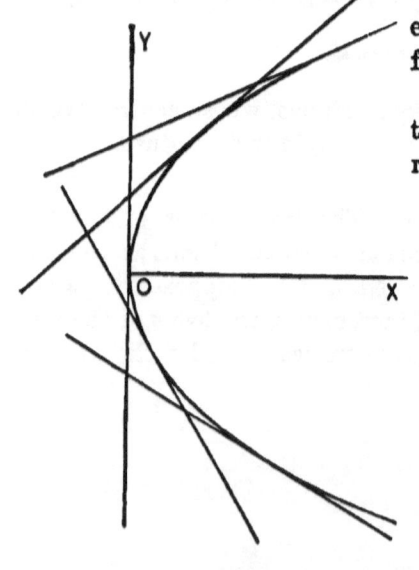

Before considering the general problem let us take the following special example.

Required the envelope of the series of straight lines represented by

$$y = ax + \frac{m}{a},$$

a being the variable parameter.

Let the equations of any two of these lines be

$$y = ax + \frac{m}{a}, \quad \ldots \quad (1)$$

and

$$y = (a+h)x + \frac{m}{a+h}. \quad (2)$$

From (1) and (2) as simultaneous equations, we can find the intersection of the two lines. Subtracting (1) from (2),

$$0 = hx - \frac{hm}{a(a+h)},$$

or
$$0 = x - \frac{m}{a(a+h)}. \quad \ldots \ldots \ldots \quad (3)$$

From (3) and (1), we have

$$x = \frac{m}{a(a+h)}, \quad y = \frac{(2a+h)m}{a(a+h)}, \quad \ldots \quad (4)$$

which are the co-ordinates of the intersection.

Now if we suppose h to approach zero in (4), we have for the ultimate intersection of consecutive lines

$$x = \frac{m}{a^2}, \quad y = \frac{2m}{a}.$$

By eliminating a between these equations we have

$$y^2 = 4mx,$$

which, being independent of a, is the equation of the locus of the intersection of *any* two consecutive lines; that is, the equation of the required envelope.

The figure shows the straight lines, and the envelope which is a parabola.

135. We will now give the general solution.

Let the given equation be

$$f(x, y, a) = 0,$$

which, by varying the parameter a, represents the series of curves.

To find the intersection of any two curves of the series, we combine
$$f(x, y, a) = 0, \quad \ldots \ldots \ldots \ldots \quad (1)$$
and
$$f(x, y, a+h) = 0. \quad \ldots \ldots \ldots \quad (2)$$

From (1) and (2), we have

$$\frac{f(x, y, a+h) - f(x, y, a)}{h} = 0, \quad \ldots \quad (3)$$

and it is evident that the intersection may be found by combining (1) and (3), instead of (1) and (2).

When the two curves approach coincidence, h approaches zero, and we have, by Art. 10, for the limit of equation (3),

$$\frac{\partial}{\partial a} f(x, y, a) = 0. \quad \ldots \quad \ldots \quad (4)$$

Thus equations (1) and (4) determine the intersection of two consecutive curves. By eliminating a between (1) and (4) we shall obtain the equation of the locus of these ultimate intersections, which is the equation of the envelope.

136. Applying this method to the preceding example,

$$y = ax + \frac{m}{a},$$

we differentiate with reference to a, and obtain for (4) Art. 135,

$$0 = x - \frac{m}{a^2}.$$

Eliminating a between these equations gives the equation of the envelope,

$$y^2 = 4mx, \quad \text{as before.}$$

137. *The evolute of a given curve is the envelope of its normals.*

This is indicated by the figure of Art. 117, and the proposition may be proved by the method of Art. 135, as follows:

The general equation of the normal at the point (x', y') is by (2) Art. 99,

$$x - x' + \frac{dy'}{dx'}(y - y') = 0, \quad \ldots \quad \ldots \quad (1)$$

in which the variable parameter is x', the quantities y', $\dfrac{dy'}{dx'}$, being functions of x'. Differentiating (1) with reference to x', we have

$$-1-\left(\frac{dy'}{dx'}\right)^2+(y-y')\frac{d^2y'}{dx'^2}=0. \quad \ldots \quad (2)$$

From (1) and (2) we find for the intersection of consecutive normals,

$$y = y' + \frac{1+\left(\dfrac{dy'}{dx'}\right)^2}{\dfrac{d^2y'}{dx'^2}},$$

$$x = x' - \frac{\dfrac{dy'}{dx'}\left[1+\left(\dfrac{dy'}{dx'}\right)^2\right]}{\dfrac{d^2y'}{dx'^2}}.$$

As these expressions are identical with the co-ordinates of the centre of curvature in Art. 116, it follows that the envelope of the normals coincides with the evolute.

EXAMPLES.

1. Find the envelope of the series of straight lines represented by $y = 2mx + m^4$, m being the variable parameter.

    Differentiating the given equation with reference to m,

    $$0 = 2x + 4m^3.$$

    Eliminating m between the two equations, we have for the envelope,
    $$16y^3 + 27x^4 = 0.$$

2. Find the envelope of the series of parabolas $y^2 = a(x-a)$, a being the variable parameter. *Ans.* $4y^2 = x^2$.

3. Find the envelope of a series of circles whose centres are on the axis of X, and radii proportional to (m times) their distance from the origin. *Ans.* $y^2 = m^2(x^2 + y^2)$.

4. Find the evolute of the parabola $y^2 = 4ax$ according to Art. 137, taking the equation of the normal in the form
$$y = m(x - 2a) - am^3. \quad \textit{Ans. } 27ay^2 = 4(x - 2a)^3.$$

5. Find the evolute of the ellipse $\dfrac{x^2}{a^2} + \dfrac{y^2}{b^2} = 1$, taking the equation of the normal in the form
$$by = ax\tan\phi - (a^2 - b^2)\sin\phi,$$
where ϕ is the eccentric angle.
$$\textit{Ans. } (ax)^{\frac{2}{3}} + (by)^{\frac{2}{3}} = (a^2 - b^2)^{\frac{2}{3}}.$$

6. Find the envelope of the straight lines represented by
$$x\cos 3\theta + y\sin 3\theta = a(\cos 2\theta)^{\frac{3}{2}},$$
θ being the variable parameter.
$$\textit{Ans. } (x^2 + y^2)^2 = a^2(x^2 - y^2), \text{ the lemniscate.}$$

7. Find the envelope of the series of ellipses, whose axes coincide and whose area is constant.

The equation of the ellipses is
$$\frac{x^2}{a^2} + \frac{y^2}{b^2} = 1, \quad\ldots\ldots\ldots\ldots \quad (1)$$

a and b being variable parameters, subject to the condition
$$ab = k^2, \quad\ldots\ldots\ldots\ldots\ldots \quad (2)$$
calling the constant area πk^2.

Substituting in (1) the value of b from (2),
$$\frac{x^2}{a^2} + \frac{a^2 y^2}{k^4} = 1, \quad\ldots\ldots\ldots\ldots \quad (3)$$

in which a is the only variable parameter. Differentiating (3) with reference to a, we have

ENVELOPES.

$$-\frac{2x^2}{a^3} + \frac{2ay^2}{k^4} = 0. \quad \ldots \ldots \quad (4)$$

Eliminating a between (3) and (4), we have

$$4x^2y^2 = k^4.$$

Second Solution. Differentiate (1), regarding both a and b as variable.

$$\frac{x^2 da}{a^3} + \frac{y^2 db}{b^3} = 0. \quad \ldots \ldots \quad (5)$$

Differentiating (2) also, we have

$$bda + adb = 0. \quad \ldots \ldots \ldots \quad (6)$$

From (5) and (6), we have

$$\frac{x^2}{a^2} = \frac{y^2}{b^2} \quad \ldots \ldots \ldots \ldots \quad (7)$$

From (7) and (1),

$$\frac{x^2}{a^2} = \frac{y^2}{b^2} = \frac{1}{2}. \quad \ldots \ldots \ldots \quad (8)$$

Substituting (8) in (2),

$$4x^2y^2 = k^4.$$

8. Find the envelope of the circles whose diameters are the double ordinates of the parabola $y^2 = 4ax$.

         Ans. $y^2 = 4a(a+x)$.

9. Find the envelope of the straight lines $\frac{x}{a} + \frac{y}{b} = 1$,

 when $a^n + b^n = k^n$.

         Ans. $x^{\frac{n}{n+1}} + y^{\frac{n}{n+1}} = k^{\frac{n}{n+1}}$.

10. Find the envelope of the ellipses $\frac{x^2}{a^2} + \frac{y^2}{b^2} = 1$,

 when $a + b = k$.   *Ans.* $x^{\frac{2}{3}} + y^{\frac{2}{3}} = k^{\frac{2}{3}}$.

11. Find the envelope of the circles passing through the origin, whose centres are on the parabola $y^2 = 4ax$.

$$\text{Ans.} \quad (x+2a)y^2 + x^3 = 0.$$

12. Find the envelope of circles described on the central radii of an ellipse as diameters, the equation of the ellipse being $\dfrac{x^2}{a^2} + \dfrac{y^2}{b^2} = 1.$ Ans. $(x^2+y^2)^2 = a^2x^2 + b^2y^2.$

13. Find the envelope of the ellipses whose axes coincide, and such that the distance between the extremities of the major and minor axes is constant and equal to k.

$$\text{Ans. A square whose sides are } (x \pm y)^2 = k^2.$$

CHAPTER XVII.

SINGULAR POINTS OF CURVES.

138. The term *singular points* is applied to points of a curve having some peculiar property independent of the position of the co-ordinate axes.

We proceed to consider the different varieties of singular points.

Points of Inflexion. These have already been considered in Art. 110.

Multiple Points. These are points through which several branches of a curve pass. The figures show a double point and a triple point.

139. *To find the multiple points of a curve.* It is evident that at such a point there are several tangents, and therefore $\frac{dy}{dx}$ has more than one value.

Suppose the equation of the curve, free from radicals, to be
$$f(x, y) = 0.$$
Then by (2) Art. 67, we have
$$\frac{dy}{dx} = -\frac{\frac{\partial u}{\partial x}}{\frac{\partial u}{\partial y}}, \quad \text{where} \quad u = f(x, y).$$

Since u contains no radicals, this expression for $\dfrac{dy}{dx}$ can have but one value at any given point, unless it takes the form $\dfrac{0}{0}$; that is,

$$\frac{\partial u}{\partial x}=0, \text{ and } \frac{\partial u}{\partial y}=0. \quad \ldots \ldots \quad (1)$$

These are therefore the conditions for a multiple point.

If values of x and y which satisfy (1) also satisfy the equation of the curve
$$f(x, y) = 0,$$
we have for any such point
$$\frac{dy}{dx} = \frac{0}{0}.$$

This indeterminate form can be evaluated by the method of Art. 53.

The result of the process of evaluation will be an equation of the second, or higher, degree with respect to $\dfrac{dy}{dx}$, thus determining several values of that quantity. This will be apparent from an example.

140. Let us examine for multiple points the lemniscate
$$(x^2 + y^2)^2 = a^2(x^2 - y^2).$$

Here
$$u = (x^2 + y^2)^2 + a^2(y^2 - x^2) = 0.$$

$$\frac{\partial u}{\partial x} = 4x(x^2 + y^2) - 2a^2 x,$$

$$\frac{\partial u}{\partial y} = 4y(x^2 + y^2) + 2a^2 y.$$

Putting $\dfrac{\partial u}{\partial x} = 0$, and $\dfrac{\partial u}{\partial y} = 0$,

we find $x = 0, \; y = 0, \;$ or $\; x = \pm \dfrac{a}{\sqrt{2}}, \; y = 0.$

SINGULAR POINTS.

Of these values of x and y, $x=0$, $y=0$, alone satisfy the equation of the given curve. Let us find the value of $\frac{dy}{dx}$ for this point.

$$-\frac{dy}{dx} = \frac{\frac{\partial u}{\partial x}}{\frac{\partial u}{\partial y}} = \frac{2x^3 + 2xy^2 - a^2x}{2x^2y + 2y^3 + a^2y} = \frac{0}{0}, \text{ when } x=0, y=0.$$

Evaluating by Art. 53,

$$-\frac{dy}{dx} = \frac{6x^2 + 2y^2 + 4xy\frac{dy}{dx} - a^2}{4xy + (2x^2 + 6y^2 + a^2)\frac{dy}{dx}} = \frac{-a^2}{a^2\frac{dy}{dx}}, \text{ when } x=0, y=0.$$

Hence $\left(\frac{dy}{dx}\right)^2 = 1,$ or $\frac{dy}{dx} = \pm 1.$

The origin is a double point, the two tangents being inclined 45° to X.

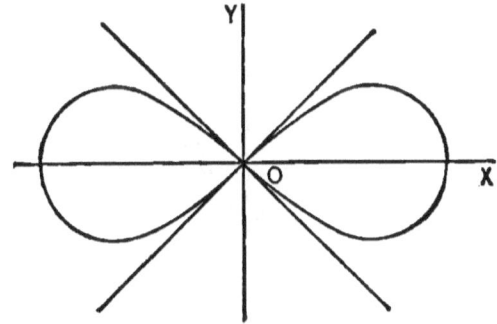

141. Again, take the curve whose equation is

$$u = x^4 + 2ax^2y - ay^3 = 0.$$

$$\frac{\partial u}{\partial x} = 4x^3 + 4axy, \quad \frac{\partial u}{\partial y} = 2ax^2 - 3ay^2.$$

Putting $\frac{\partial u}{\partial x} = 0$, and $\frac{\partial u}{\partial y} = 0$, we find $x = 0$, $y = 0$, to be the only point of the curve satisfying these conditions.

In finding the values of $\frac{dy}{dx}$,

let $\qquad y_1 = \frac{dy}{dx},\qquad$ and $\qquad y_2 = \frac{d^2y}{dx^2}.$

$$y_1 = \frac{4x^3 + 4axy}{3ay^2 - 2ax^2} = \frac{0}{0}, \text{ when } x = 0, y = 0.$$

Evaluating by Art. 53,

$$y_1 = \frac{12x^2 + 4ay + 4axy_1}{6ayy_1 - 4ax} = \frac{0}{0}, \text{ when } x = 0, y = 0.$$

Evaluating again,

$$y_1 = \frac{24x + 8ay_1 + 4axy_2}{6ay_1^2 + 6ayy_2 - 4a} = \frac{8ay_1}{6ay_1^2 - 4a}, \text{ when } x = 0, y = 0.$$

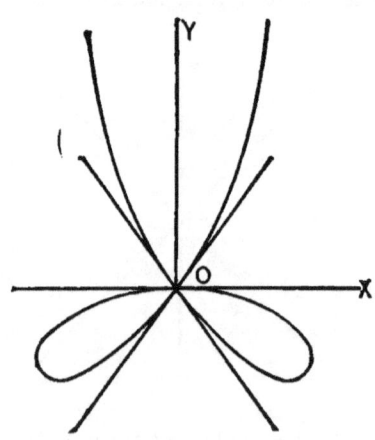

Hence $\qquad y_1(3y_1^2 - 2) = 4y_1,$

and therefore $\qquad y_1 = 0,\;$ or $\;y_1 = \pm\sqrt{2}.$

Hence the origin is a triple point as shown in the figure.

SINGULAR POINTS. 149

142. *Points of Osculation.* A multiple point is called a *point of osculation* when the branches of the curve passing through it are tangent to each other.

In this case $\dfrac{dy}{dx}$ will have two or more equal values at the point.

For example, consider the curve

$$a^4 y^2 = a^2 x^4 - x^6.$$

Here $\quad u = a^4 y^2 - a^2 x^4 + x^6 = 0.$

$$\frac{\partial u}{\partial x} = -4 a^2 x^3 + 6 x^5, \qquad \frac{\partial u}{\partial y} = 2 a^4 y.$$

$$y_1 = \frac{4 a^2 x^3 - 6 x^5}{2 a^4 y} = \frac{0}{0}, \quad \text{when } x = 0,\ y = 0.$$

Evaluating by Art. 53,

$$y_1 = \frac{12 a^2 x^2 - 30 x^4}{2 a^4 y_1} = \frac{0}{2 a^4 y_1}, \quad \text{when } x = 0,\ y = 0.$$

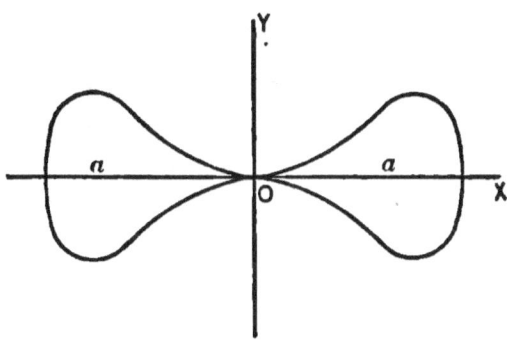

Hence $2 a^4 y_1^2 = 0$, giving two values of $y_1 = 0$. The origin is a point of osculation.

143. Cusps. When the branches of the curve are only on one side of the point of osculation, this point is called a cusp, as P_1 or P_2.

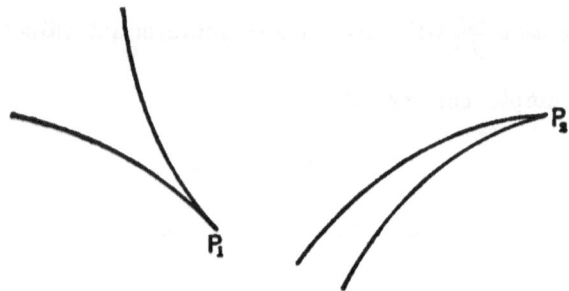

The conditions for a cusp are the same as those for a point of osculation, with the additional condition of imaginary points of the curve on one side of this point.

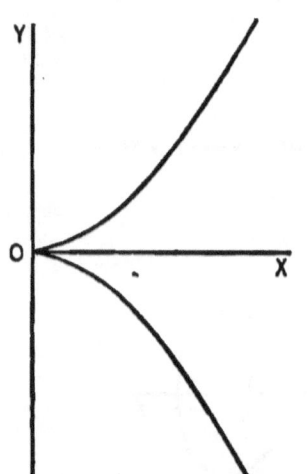

For example, take the semicubical parabola
$$y^2 = x^3.$$
Here $\quad y = \pm x^{\frac{3}{2}},$

$$\frac{dy}{dx} = \pm \frac{3}{2} x^{\frac{1}{2}}.$$

When $x = 0, \quad \dfrac{dy}{dx} = \pm\, 0.$

There are then two coincident tangents at the origin. But since y is imaginary for negative values of x, there are no points on the left of the origin. Hence the origin is a cusp.

144. Conjugate Points. If, in determining a multiple point, the values of $\dfrac{dy}{dx}$ are imaginary, we then have a point of the curve through which *no* branches pass; that is, an isolated point. Such a point is called a *conjugate point*.

SINGULAR POINTS. 151

For example, the curve

$$ay^2 - x^3 + bx^2 = 0, \quad \text{gives}$$

$$\frac{dy}{dx} = \frac{3x^2 - 2bx}{2ay} = \frac{0}{0}, \quad \text{when } x = 0, y = 0.$$

Hence

$$\frac{dy}{dx} = \frac{6x - 2b}{2a\frac{dy}{dx}} = -\frac{b}{a\frac{dy}{dx}},$$

when $x = 0, y = 0$.

Therefore

$$\frac{dy}{dx} = \pm \sqrt{-\frac{b}{a}}.$$

Hence the origin is a conjugate point. This appears directly from the given equation

$$ay^2 = x^2(x - b),$$

from which it is evident that besides the origin, there are no points of the curve when $x < b$.

EXAMPLES.

1. Show that the curve

$$a^2 y^2 = a^2 x^2 - x^4,$$

has a multiple point at the origin.

2. Show that the curve

$$y^2 = x \log(1 + x),$$

has a multiple point at the origin.

3. Show that the cissoid
$$y^2 = \frac{x^3}{2a-x},$$
has a cusp at the origin.

4. Show that the curve
$$x^3 + 2x^2 + 2xy - y^2 + 5x - 2y = 0,$$
has a cusp at the point $(-1, -2)$.

5. Show that the curve
$$(x^2 + y^2)^2 = a^2x^2 + b^2y^2,$$
has a conjugate point at the origin.

6. Show that the curve
$$ay^2 = (x-a)^2(x-b), \quad \text{at the point } (a, 0),$$
has a conjugate point, if $a < b$;
a double point, if $a > b$;
and a cusp, if $a = b$.

CHAPTER XVIII.

MAXIMA AND MINIMA OF FUNCTIONS OF ONE INDEPENDENT VARIABLE.

Jan 29, '06,

145. *Definition.* A *maximum* value of a function is a value *greater* than those immediately preceding or immediately following.

A *minimum* value of a function is a value *less* than those immediately preceding or immediately following.

If the function is represented by the curve $y = f(x)$, then

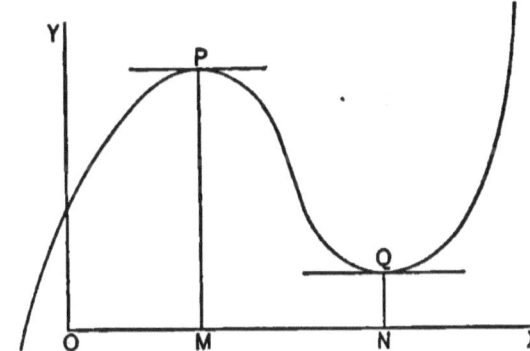

PM represents a maximum value of y or of $f(x)$, and QN represents a minimum value.

146. *To find the conditions for a maximum or a minimum.*

It is evident that at both P and Q the tangent is parallel to the axis of X, and therefore we have as a condition for both maxima and minima,

$$\frac{dy}{dx} = 0 \quad \ldots \ldots \ldots \ldots \quad (a)$$

Again, at P the curve is concave *downward*, and at Q, concave *upward*.

Hence, by Art. 109,

$$\left. \begin{array}{l} \text{for a maximum value,} \ \dfrac{d^2y}{dx^2} < 0, \\ \text{for a minimum value,} \ \dfrac{d^2y}{dx^2} > 0. \end{array} \right\} \quad \ldots \quad (b)$$

For example, find the maximum and minimum value of

$$\frac{x^3}{3} - 2x^2 + 3x + 1.$$

Put $\qquad y = \dfrac{x^3}{3} - 2x^2 + 3x + 1.$

Then $\dfrac{dy}{dx} = x^2 - 4x + 3$, $\dfrac{d^2y}{dx^2} = 2x - 4$.

By (a), $x^2 - 4x + 3 = 0$.

Solving this equation,

$x = 1$ or 3.

To apply (b), we substitute both $x = 1$ and $x = 3$ in

$$\dfrac{d^2y}{dx^2} = 2x - 4,$$

and find when $x = 1$, $\dfrac{d^2y}{dx^2} < 0$,

when $x = 3$, $\dfrac{d^2y}{dx^2} > 0$.

Hence when $x = 1$, y is a maximum;

when $x = 3$, y is a minimum.

The maximum value of y is $2\frac{1}{3}$, and the minimum value, 1.

147. In exceptional cases it may happen that the value of x given by (a) makes $\dfrac{d^2y}{dx^2} = 0$, so that neither of the con-

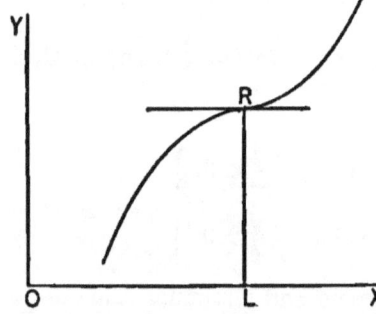

ditions (b) is satisfied. This would be the case for a point of inflexion R, whose tangent is parallel to OX. Here the ordinate RL is neither a maximum nor a minimum.

But there may be a maximum or minimum value of y, even when $\dfrac{d^2y}{dx^2} = 0$. This is more fully considered in Art. 150. The method of the following article is also applicable to such cases.

148. *Second Method of determining Maxima and Minima.*

Referring to the figure of Art. 145, and supposing x to

MAXIMA AND MINIMA FOR ONE VARIABLE. 155

increase, we see that as we approach P, y increases, and on leaving P, y decreases. Hence, by Art. 108, $\frac{dy}{dx}$ is positive on the left, and negative on the right, of P. That is, when y is a maximum, $\frac{dy}{dx}$ changes from $+$ to $-$.

Similarly, it may be shown that when, as at Q, y is a minimum, $\frac{dy}{dx}$ changes from $-$ to $+$.

These relations may also be obtained by noticing that $\tan \phi$, which is equal to $\frac{dy}{dx}$, changes sign at P and Q.

Let us apply these conditions to the example in Art. 146, where

$$\frac{dy}{dx} = x^2 - 4x + 3 = (x-1)(x-3).$$

Here $\frac{dy}{dx}$ can change sign only when $x = 1$ or $x = 3$.

By supposing x to be first slightly less, and then slightly greater, than 1, we find that $x - 1$ changes from $-$ to $+$; but since $x - 3$ is then negative, it follows that $\frac{dy}{dx}$ changes from $+$ to $-$, when $x = 1$, and denotes a maximum. In the same way, we find that $\frac{dy}{dx}$ changes from $-$ to $+$, when $x = 3$, and denotes a minimum.

Again, consider the function $y = (x-4)^5(x+2)^4$.

Here $\quad \frac{dy}{dx} = 3(3x-2)(x-4)^4(x+2)^3.$

When $x = \frac{2}{3}$, $\frac{dy}{dx}$ changes from $-$ to $+$;

when $x = -2$, $\frac{dy}{dx}$ changes from $+$ to $-$;

when $x = 4$, $\frac{dy}{dx}$ does *not* change sign,

since $(x-4)^4$ cannot be negative.

Hence we conclude that y is a minimum when $x = \frac{2}{3}$; a maximum when $x = -2$; but neither a maximum nor minimum when $x = 4$.

As this method does not require $\frac{d^2y}{dx^2}$, it is preferable to that of Art. 146, when the second differentiation of y involves much work.

149. *Case where $\frac{dy}{dx} = \infty$.* It is to be noticed that $\frac{dy}{dx}$ sometimes changes sign by passing through infinity instead of zero.

Hence if
$$\frac{dy}{dx} = \infty,$$
for a finite value of x, this value should be examined, as well as those given by
$$\frac{dy}{dx} = 0.$$

For example, suppose
$$y = a - b(x - c)^{\frac{2}{3}}.$$

Then
$$\frac{dy}{dx} = -\frac{2b}{3(x-c)^{\frac{1}{3}}};$$

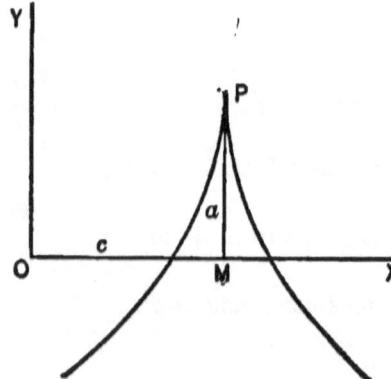

hence we have
$$\frac{dy}{dx} = \infty, \quad \text{when } x = c.$$

It is evident that when $x = c$, $\frac{dy}{dx}$ changes from $+$ to $-$, indicating a maximum value of y, which is a.

The figure shows the maximum ordinate PM, corresponding to a cusp at P.

MAXIMA AND MINIMA FOR ONE VARIABLE. 157

On the other hand, suppose $y = a - b(x - c)^{\frac{1}{3}}$.

Then $\dfrac{dy}{dx} = -\dfrac{b}{3(x-c)^{\frac{2}{3}}} = \infty$, when $x = c$.

But as $\dfrac{dy}{dx}$ does not change sign when $x = c$, there is no maximum nor minimum. The corresponding curve is shown in the figure.

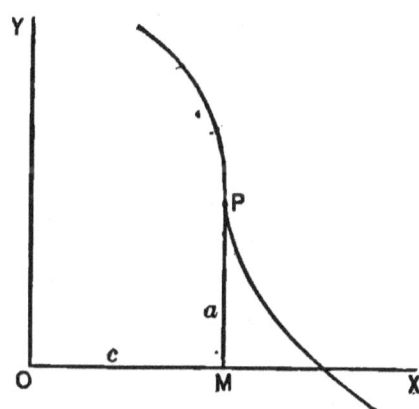

150. Conditions for Maxima and Minima by Taylor's Theorem. Suppose the function $f(x)$ to be a maximum when $x = a$. Then, by the definition in Art. 145,

$$f(a) > f(a + h),$$

and also $\qquad f(a) > f(a - h),$

where h is any small but finite quantity. Now, by the substitution of a for x in Taylor's Theorem, we have

$$f(a+h) - f(a) = hf'(a) + \frac{h^2}{\underline{|2}}f''(a) + \frac{h^3}{\underline{|3}}f'''(a) + \cdots \quad (1)$$

$$f(a-h) - f(a) = -hf'(a) + \frac{h^2}{\underline{|2}}f''(a) - \frac{h^3}{\underline{|3}}f'''(a) + \cdots \quad (2)$$

By the hypothesis $\quad f(a+h) - f(a) < 0,$

and also $\qquad f(a-h) - f(a) < 0.$

Hence the second members of both (1) and (2) must be negative.

By taking h sufficiently small, the first term can be made numerically greater than the sum of all the others, involving h^2, h^3, etc. Thus the sign of the entire second member will be that of the first term. As these have *different* signs in (1) and (2), the second members cannot both be negative unless

$$f'(a) = 0.$$

Equations (1) and (2) then become

$$f(a+h) - f(a) = \frac{h^2}{\lfloor 2}f''(a) + \frac{h^3}{\lfloor 3}f'''(a) + \cdots$$

$$f(a-h) - f(a) = \frac{h^2}{\lfloor 2}f''(a) - \frac{h^3}{\lfloor 3}f'''(a) + \cdots.$$

The term containing h^2 now determines the sign of the second members. That these may be negative, we must have

$$f''(a) < 0.$$

If then $\qquad f'(a) = 0 \quad \text{and} \quad f''(a) < 0,$

$f(a)$ is a maximum.

Similarly, it may be shown that if

$$f'(a) = 0 \quad \text{and} \quad f''(a) > 0,$$

$f(a)$ will be a minimum.

If $\qquad f'(a) = 0 \quad \text{and} \quad f''(a) = 0,$

similar reasoning will show that for a maximum we must also have
$$f'''(a) = 0 \quad \text{and} \quad f^{\text{iv}}(a) < 0;$$

and for a minimum
$$f'''(a) = 0 \quad \text{and} \quad f^{\text{iv}}(a) > 0.$$

151. The conditions may be generalized as follows:
Suppose

$$f'(a) = 0, \quad f''(a) = 0, \quad f'''(a) = 0, \quad \cdots \quad f^n(a) = 0.$$

Then if n is even, $f(a)$ is neither a maximum nor a minimum.

If n is odd, $f(a)$ will be a maximum or minimum, according as
$$f^{n+1}(a) < 0 \quad \text{or} \quad > 0.$$

EXAMPLES.

1. Find the maximum value of $ax - x^2$. Ans. $\dfrac{a^2}{4}$, when $x = \dfrac{a}{2}$.

2. Find the maximum and minimum values of
$2x^3 - 9x^2 + 12x - 3$. Ans. $x = 1$ gives a maximum, 2;
$x = 2$ gives a minimum, 1.

3. Find the maximum and minimum values of
$x^3 - 3x^2 - 9x + 5$. Ans. $x = -1$ gives a maximum, 10;
$x = 3$ gives a minimum, -22.

4. Show that $x^3 - 3x^2 + 6x$ has neither a maximum nor minimum value.

5. Show that $ax + \dfrac{b}{x}$, is a minimum, when $ax = \dfrac{b}{x} = \sqrt{ab}$.

6. Show that the least value of $\dfrac{a^2}{\sin^2\theta} + \dfrac{b^2}{\cos^2\theta}$ is $(a+b)^2$.

Investigate the following functions for maxima or minima:

7. $y = \dfrac{x^2 - 7x + 6}{x - 10}$. Ans. $x = 4$ gives a maximum value of y; $x = 16$ gives a minimum value of y.

8. $y = \dfrac{x}{\log x}$. Ans. A minimum when $x = e$.

9. $y = \dfrac{(x-a)(b-x)}{x^2}$.

Ans. $x = \dfrac{2ab}{a+b}$ gives a maximum value, $\dfrac{(a-b)^2}{4ab}$.

10. $y = 2\tan x - \tan^2 x$. Ans. A maximum when $x = \dfrac{\pi}{4}$.

11. $y = \sin x(1 + \cos x)$. *Ans.* A maximum when $x = \dfrac{\pi}{3}$.

12. $y = \tan x + 3 \cot x$. *Ans.* A minimum when $x = \dfrac{\pi}{3}$.

13. $y = \sin x \cos(x - a)$. *Ans.* A maximum when $x = \dfrac{a}{2} + \dfrac{\pi}{4}$,

   a minimum when $x = \dfrac{a}{2} - \dfrac{\pi}{4}$.

14. $y = \dfrac{(a-x)^3}{a-2x}$. *Ans.* A minimum when $x = \dfrac{a}{4}$.

15. $y = (x-1)^4(x+2)^3$.
   Ans. A maximum when $x = -\dfrac{5}{7}$; a minimum when $x = 1$; neither when $x = -2$.

16. $y = (x-2)^5(2x+1)^4$.
   Ans. A maximum when $x = -\dfrac{1}{2}$; a minimum when $x = \dfrac{11}{18}$; neither when $x = 2$.

17. $y = (x+1)^{\frac{2}{3}}(x-5)^2$.
   Ans. A minimum when $x = 5$; a maximum when $x = \dfrac{1}{2}$; a minimum when $x = -1$.

18. $y = (2x-a)^{\frac{1}{3}}(x-a)^{\frac{2}{3}}$.
   Ans. A maximum when $x = \dfrac{2a}{3}$; a minimum when $x = a$.

PROBLEMS IN MAXIMA AND MINIMA.

1. Divide 10 into two such parts that the product of the square of one and the cube of the other may be the greatest possible.

Let x and $10-x$ be the parts. Then $x^2(10-x)^3$ is to be a maximum. Letting $u = x^2(10-x)^3$, we find

$$\frac{du}{dx} = 5x(4-x)(10-x)^2 = 0,$$

from which we find that u is a maximum when $x = 4$. Hence the required parts are 4 and 6.

MAXIMA AND MINIMA FOR ONE VARIABLE. 161

2. A square piece of pasteboard whose side is a, has a small square cut out at each corner; find the side of this square that the remainder may form a box of maximum contents.

Let $x =$ the side of the small square. Then the contents of the box will be $(a-2x)^2 x$. Representing this by u, we find that u is a maximum when $x = \dfrac{a}{6}$, which is the required answer.

3. Find the greatest right cylinder that can be inscribed in a given right cone.

Suppose the figure to be a section through the axis AD.

Let $AD = a$, $DC = b$.

Let $x = DQ$, the radius of the base of the cylinder, and $y = PQ$, its altitude.

From the similar triangles ADC, PQC, we find

$$\frac{y}{b-x} = \frac{a}{b} \quad \text{or} \quad y = \frac{a}{b}(b-x).$$

The volume of the cylinder is

$$\pi x^2 y = \pi \frac{a}{b} x^2 (b-x).$$

This will be a maximum when $u = bx^2 - x^3$ is a maximum.

This is found to be when $x = \tfrac{2}{3}b$, the radius of the base of the required cylinder.

From this, $y = \dfrac{a}{3}$, the altitude of the cylinder.

4. Determine the right cylinder of the greatest convex surface that can be inscribed in a given sphere.

Suppose the figure (page 162) to be a section through the axis of the cylinder, AB.

Let $r = OP$, the radius of the sphere.

Let $x = OR$, the radius of the base of the cylinder, and $y = PR$, one-half its altitude.

From the right triangle OPR we have

$$x^2 + y^2 = r^2. \quad \ldots \ldots \ldots \ldots \quad (a)$$

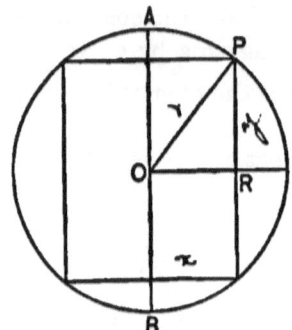

The convex surface of the cylinder is

$$2\pi x \cdot 2y = 4\pi x \sqrt{r^2 - x^2}$$
$$= 4\pi \sqrt{r^2 x^2 - x^4}.$$

This will be a maximum when $u = r^2 x^2 - x^4$ is a maximum.

This is found to be when $x = \dfrac{r}{\sqrt{2}}$, the radius of the base of the required cylinder.

From this, $y = \dfrac{r}{\sqrt{2}}$. Hence the altitude of the cylinder is $r\sqrt{2}$.

Another solution of the problem is the following:

Since the convex surface is $4\pi xy$, put $u = xy$, to be a maximum.

$$\frac{du}{dx} = y + x\frac{dy}{dx} = 0. \quad \ldots \ldots \ldots \quad (b)$$

But from (a), $\qquad x + y\dfrac{dy}{dx} = 0. \quad \ldots \ldots \ldots \quad (c)$

Eliminating $\dfrac{dy}{dx}$ from (b) and (c), we have $x = y$, which, combined with (a), gives the same result as before.

5. A rectangular piece of pasteboard 30 inches long and 14 inches wide has a square cut out at each corner; find the side of this square so that the remainder may form a box of maximum contents. *Ans.* 3 inches.

6. Divide a into two parts such that the product of the mth power of one and the nth power of the other may be a maximum. *Ans.* The required parts are proportional to m and n.

7. A person being in a boat 3 miles from the nearest point of the beach, wishes to reach in the shortest time a place 5 miles from that point along the shore; supposing he can walk 5 miles an hour, but row only at the rate of 4 miles an hour, required the place he must land.

Ans. One mile from the place to be reached.

8. The top of a pedestal which sustains a statue 11 feet high is 25 feet above the level of a man's eye; find his horizontal distance from the base of the pedestal when he sees the statue subtending the greatest angle. *Ans.* 30 feet.

9. Through a point (a, b), referred to rectangular axes, a straight line is to be drawn, forming with the axes a triangle of the least area. Show that its intercepts on the axes are $2a$ and $2b$.

10. Through the point (a, b) a line is drawn such that the part intercepted between the axes is a minimum. Show that its length is $(a^{\frac{2}{3}} + b^{\frac{2}{3}})^{\frac{3}{2}}$.

11. Given the slant height a of a right cone; find its altitude when the volume is a maximum.

Ans. $\dfrac{a}{\sqrt{3}}$.

12. Given a point on the axis of the parabola $y^2 = 4ax$, at the distance h from the vertex; find the abscissa of the point of the curve nearest to it. *Ans.* $x = h - 2a$.

13. Find the maximum rectangle that can be inscribed in an ellipse whose semi-axes are a and b.

Ans. The sides are $a\sqrt{2}$ and $b\sqrt{2}$; the area, $2ab$.

14. A rectangular box, open at the top, with a square base, is to be constructed to contain 108 cubic inches. What must be its dimensions to require the least material?

Ans. Altitude, 3 inches; side of base, 6 inches.

15. Find the altitude of the right cylinder of greatest volume inscribed in a sphere whose radius is r. *Ans.* $\dfrac{2r}{\sqrt{3}}$.

16. Find the altitude of the right cylinder inscribed in a sphere whose radius is r, when its entire surface is a maximum.

$$Ans.\ \left(2-\frac{2}{\sqrt{5}}\right)^{\frac{1}{2}}r.$$

17. Find the altitude of the right cone of greatest volume inscribed in a sphere whose radius is r. $\quad Ans.\ \tfrac{4}{3}r.$

18. Find the altitude of the right cone of maximum entire surface inscribed in a sphere whose radius is r.

$$Ans.\ (23-\sqrt{17})\frac{r}{16}.$$

19. Find the altitude of the right cone of least volume circumscribed about a sphere whose radius is r.

Ans. Its altitude is $4r$, and its volume is twice that of the sphere.

20. Find the altitude of the least isosceles triangle circumscribed about an ellipse whose semi-axes are a and b, the base of the triangle being parallel to the major axis.

$$Ans.\ 3b.$$

21. A tangent is drawn to the ellipse whose semi-axes are a and b, such that the part intercepted by the axes is a minimum. Show that its length is $a+b$.

22. The lower corner of a leaf, whose width is a, is folded over so as just to reach the inner edge of the page. Find the width of the part folded over, when the length of the crease is a minimum. $\quad Ans.\ \tfrac{3}{4}a.$

23. In the preceding example, find when the area of the triangle folded over is a minimum.

$\quad Ans.$ When the width folded is $\tfrac{2}{3}a.$

CHAPTER XIX.

MAXIMA AND MINIMA OF FUNCTIONS OF TWO OR MORE INDEPENDENT VARIABLES.

152. *Definition.* A function, $f(x, y)$, of two independent variables has a *maximum* value, when

$$f(x, y) > f(x + h, y + k),$$

for all small values of h and k, positive or negative; and a *minimum* value, when

$$f(x, y) < f(x + h, y + k).$$

153. *Conditions for Maxima or Minima.*

Letting $\qquad u = f(x, y),$

we have from Art. 68,

$$f(x + h, y + k) - f(x, y) = h\frac{\partial u}{\partial x} + k\frac{\partial u}{\partial y}$$
$$+ \frac{1}{2}\left(h^2\frac{\partial^2 u}{\partial x^2} + 2hk\frac{\partial^2 u}{\partial x \partial y} + k^2\frac{\partial^2 u}{\partial y^2}\right) + \ \cdot \ \cdot \ \cdot \quad (1)$$

In order that u may be a maximum, the second member of (1) must be negative for small values of h and k, positive or negative. By similar reasoning to that in Art. 150, it is evident that the sign of (1) is determined by the terms containing the lowest powers of h and k; that is, by

$$h\frac{\partial u}{\partial x} + k\frac{\partial u}{\partial y}.$$

Hence, in order that (1) may not change sign with h and k, we must have

$$h\frac{\partial u}{\partial x} + k\frac{\partial u}{\partial y} = 0.$$

As h and k are independent of each other, this is equivalent to

$$\frac{\partial u}{\partial x} = 0, \quad \text{and} \quad \frac{\partial u}{\partial y} = 0. \quad \ldots \ldots (2)$$

Equation (1) then becomes

$$f(x+h, y+k) - f(x, y) = \frac{1}{2}(Ah^2 + 2Bhk + Ck^2) + \cdots,$$

where
$$A = \frac{\partial^2 u}{\partial x^2}, \quad B = \frac{\partial^2 u}{\partial x \partial y}, \quad C = \frac{\partial^2 u}{\partial y^2}.$$

But $\quad Ah^2 + 2Bhk + Ck^2 = \dfrac{(Ah+Bk)^2 + (AC-B^2)k^2}{A}. \quad . \quad (3)$

In order that (3) may preserve the same sign for all small values of h and k, it is necessary that $AC - B^2$ should be positive; for if negative, the numerator of (3) will be positive when $k = 0$, and negative when $Ah + Bk = 0$. Hence we have as an additional condition for a maximum,

$$B^2 < AC. \quad \ldots \ldots (4)$$

The sign of (3) then depends upon that of the denominator A. Hence for a maximum we must have

$$A < 0.$$

Similarly it may be shown that for a minimum value of u, we must have (2) and (4), together with

$$A > 0.$$

It may be noticed that (4) requires that A and C should have the same sign. Hence if A is positive, C will be also.

The exceptional cases, where

$$B^2 = AC,$$

or where $\quad A = 0, \quad B = 0, \quad C = 0,$

require further investigation. We shall not consider them here.

154. The conditions for a maximum or minimum value of $u = f(x, y)$, may be restated as follows:

For either a maximum or minimum,

$$\frac{\partial u}{\partial x} = 0, \quad \text{and} \quad \frac{\partial u}{\partial y} = 0; \quad \ldots \quad (1)$$

also
$$\left(\frac{\partial^2 u}{\partial y \partial x}\right)^2 < \frac{\partial^2 u}{\partial x^2} \frac{\partial^2 u}{\partial y^2}. \quad \ldots \quad (2)$$

For a maximum,
$$\frac{\partial^2 u}{\partial x^2} < 0, \quad \text{and} \quad \frac{\partial^2 u}{\partial y^2} < 0. \quad \ldots \quad (3)$$

For a minimum,
$$\frac{\partial^2 u}{\partial x^2} > 0, \quad \text{and} \quad \frac{\partial^2 u}{\partial y^2} > 0. \quad \ldots \quad (4)$$

155. *Functions of Three Variables.* A similar investigation to that in Art. 153, gives as the conditions of a maximum or minimum value of $u = f(x, y, z)$:—

For either a maximum or minimum,

$$\frac{\partial u}{\partial x} = 0, \quad \frac{\partial u}{\partial y} = 0, \quad \frac{\partial u}{\partial z} = 0,$$

and
$$\left(\frac{\partial^2 u}{\partial x \partial y}\right)^2 < \frac{\partial^2 u}{\partial x^2} \frac{\partial^2 u}{\partial y^2}.$$

For a maximum $\quad \dfrac{\partial^2 u}{\partial x^2} < 0, \quad$ and $\quad \Delta < 0;$

for a minimum, $\quad \dfrac{\partial^2 u}{\partial x^2} > 0, \quad$ and $\quad \Delta > 0;$

where
$$\Delta = \begin{vmatrix} \dfrac{\partial^2 u}{\partial x^2}, & \dfrac{\partial^2 u}{\partial x \partial y}, & \dfrac{\partial^2 u}{\partial x \partial z} \\ \dfrac{\partial^2 u}{\partial x \partial y}, & \dfrac{\partial^2 u}{\partial y^2}, & \dfrac{\partial^2 u}{\partial y \partial z} \\ \dfrac{\partial^2 u}{\partial x \partial z}, & \dfrac{\partial^2 u}{\partial y \partial z}, & \dfrac{\partial^2 u}{\partial z^2} \end{vmatrix}.$$

EXAMPLES.

1. Find the maximum value of
$$u = 3axy - x^3 - y^3.$$

Here $\dfrac{\partial u}{\partial x} = 3ay - 3x^2$, $\dfrac{\partial u}{\partial y} = 3ax - 3y^2$.

Also $\dfrac{\partial^2 u}{\partial x^2} = -6x$, $\dfrac{\partial^2 u}{\partial y^2} = -6y$, $\dfrac{\partial^2 u}{\partial x \partial y} = 3a.$

Applying (1) Art. 154, we have
$$ay - x^2 = 0, \quad \text{and} \quad ax - y^2 = 0;$$
whence $x = 0, y = 0$; or $x = a, y = a$.

The values $x = 0, y = 0$, give
$$\dfrac{\partial^2 u}{\partial x^2} = 0, \quad \dfrac{\partial^2 u}{\partial y^2} = 0, \quad \dfrac{\partial^2 u}{\partial x \partial y} = 3a,$$

which do not satisfy (2) Art. 154.
Hence they do not give a maximum or minimum.
The values $x = a, y = a$, give
$$\dfrac{\partial^2 u}{\partial x^2} = -6a, \quad \dfrac{\partial^2 u}{\partial y^2} = -6a, \quad \dfrac{\partial^2 u}{\partial x \partial y} = 3a,$$

which satisfy both (2) and (3), Art. 154.
Hence they give a maximum value of u which is a^3.

2. Find the maximum value of xyz, subject to the condition
$$\dfrac{x^2}{a^2} + \dfrac{y^2}{b^2} + \dfrac{z^2}{c^2} = 1. \quad \ldots \ldots \ldots \quad (1)$$

From (1), $\dfrac{z^2}{c^2} = 1 - \dfrac{x^2}{a^2} - \dfrac{y^2}{b^2};$

and as xyz is numerically a maximum when $x^2 y^2 z^2$ is a maximum, we put
$$u = x^2 y^2 \left(1 - \dfrac{x^2}{a^2} - \dfrac{y^2}{b^2}\right).$$

$$\frac{\partial u}{\partial x} = 2xy^2\left(1 - \frac{2x^2}{a^2} - \frac{y^2}{b^2}\right), \quad \frac{\partial u}{\partial y} = 2x^2 y\left(1 - \frac{x^2}{a^2} - \frac{2y^2}{b^2}\right),$$

$$\frac{\partial^2 u}{\partial x^2} = 2y^2\left(1 - \frac{6x^2}{a^2} - \frac{y^2}{b^2}\right), \quad \frac{\partial^2 u}{\partial y^2} = 2x^2\left(1 - \frac{x^2}{a^2} - \frac{6y^2}{b^2}\right),$$

$$\frac{\partial^2 u}{\partial x \partial y} = 4xy\left(1 - \frac{2x^2}{a^2} - \frac{2y^2}{b^2}\right).$$

From $\dfrac{\partial u}{\partial x} = 0$ and $\dfrac{\partial u}{\partial y} = 0$, we find, as the only values satisfying (2) Art. 154,

$$x = \frac{a}{\sqrt{3}}, \quad y = \frac{b}{\sqrt{3}}, \quad \text{which give}$$

$$\frac{\partial^2 u}{\partial x^2} = -\frac{8b^2}{9}, \quad \frac{\partial^2 u}{\partial y^2} = -\frac{8a^2}{9}, \quad \frac{\partial^2 u}{\partial x \partial y} = -\frac{4ab}{9}.$$

As these values satisfy (2) and (3), Art. 154, it follows that xyz is a maximum when

$$x = \frac{a}{\sqrt{3}}, \quad y = \frac{b}{\sqrt{3}}, \quad z = \frac{c}{\sqrt{3}}.$$

The maximum value of xyz is $\dfrac{abc}{3\sqrt{3}}$.

3. Find the values of x, y, z, that render

$$x^2 + y^2 + z^2 + x - 2z - xy$$

a minimum. *Ans.* $x = -\dfrac{2}{3}, \ y = -\dfrac{1}{3}, \ z = 1.$

4. Find the maximum value of

$$(a-x)(a-y)(x+y-a). \quad Ans. \ \frac{a^3}{27}.$$

5. Find the minimum value of

$$x^2 + xy + y^2 - ax - by.$$

Ans. $\dfrac{1}{3}(ab - a^2 - b^2).$

6. Find the values of x and y that render
$$\sin x + \sin y + \cos(x+y)$$
a maximum or minimum.

Ans. A minimum, when $x = y = \dfrac{3\pi}{2}$;

a maximum, when $x = y = \dfrac{\pi}{6}$.

7. Find the maximum value of
$$\frac{(ax + by + c)^2}{x^2 + y^2 + 1}. \qquad \textit{Ans. } a^2 + b^2 + c^2.$$

8. Find the maximum value of $x^2 y^3 z^4$, subject to the condition
$$2x + 3y + 4z = a. \qquad \textit{Ans. } \left(\frac{a}{9}\right)^9.$$

9. Divide a into three parts such that their continued product may be the greatest possible.

Let the parts be x, y, and $a - x - y$.

Then $u = xy(a - x - y)$, to be a maximum.

$$\frac{\partial u}{\partial x} = ay - 2xy - y^2 = 0, \quad \frac{\partial u}{\partial y} = ax - x^2 - 2xy = 0.$$

These equations give $x = y = \dfrac{a}{3}$.

Hence a is divided into equal parts.

NOTE.—When, from the nature of the problem, it is evident that there is a maximum or minimum, it is often unnecessary to consider the second differential coefficients.

10. Divide a into three parts, x, y, z, such that $x^m y^n z^p$ may be a maximum.

$$\textit{Ans. } \frac{x}{m} = \frac{y}{n} = \frac{z}{p} = \frac{a}{m+n+p}.$$

MAXIMA AND MINIMA FOR TWO VARIABLES. 171

11. Divide 30 into four parts such that the continued product of the first, the square of the second, the cube the third, and the fourth power of the fourth, may be a maximum.

    Ans. 3, 6, 9, 12.

12. Given the volume a^3 of a rectangular parallelopiped; find when the surface is a minimum.

    Ans. When the parallelopiped is a cube.

13. An open vessel is to be constructed in the form of a rectangular parallelopiped, capable of containing 108 cubic inches of water. What must be its dimensions to require the least material in construction?

    Ans. Length and width, 6 in.; height, 3 in.

14. Find the co-ordinates of a point, the sum of the squares of whose distances from three given points,

    $$(x_1, y_1), (x_2, y_2), (x_3, y_3),$$

    is a minimum. *Ans.* $\frac{1}{3}(x_1 + x_2 + x_3)$, $\frac{1}{3}(y_1 + y_2 + y_3)$,

    the centre of gravity of the triangle joining the given points.

15. Find the volume of the greatest rectangular parallelopiped that can be inscribed in the ellipsoid

    $$\frac{x^2}{a^2} + \frac{y^2}{b^2} + \frac{z^2}{c^2} = 1.$$

    Ans. $\dfrac{8abc}{3\sqrt{3}}$.

INTEGRAL CALCULUS.

CHAPTER I.

ELEMENTARY FORMS OF INTEGRATION.

1. *Definition of Integration.* The inverse operation of differentiation is called *integration*. By differentiation we find the differential of a given function, and by integration we find the function corresponding to a given differential. This function is called the *integral* of the differential.

For instance,

since $\quad\quad 2x\,dx$ is the differential of x^2,

therefore $\quad x^2$ is the integral of $2x\,dx$.

The symbol $\int$ is used to denote the integral of the expression following it.

Thus the foregoing relations would be written,

$$d(x^2) = 2x\,dx, \quad \int 2x\,dx = x^2.$$

It is evidently the same thing whether we consider this integral, as the function whose differential is $2x\,dx$, or the function whose differential coefficient is $2x$.

As a matter of notation, however, it is not customary to write

$$\int 2x = x^2, \quad \text{but always} \quad \int 2x\,dx = x^2.$$

Integration is not like differentiation a direct operation, but consists in recognizing the given expression as the differential of a known function, or in reducing it to a form where such recognition is possible. All functions can be differentiated, but all cannot be integrated; that is, their integrals cannot be expressed in terms of known functions.

2. *Elementary Principles.*

(*a*). It is evident that we have

$$\int 2x\,dx = x^2 + 2, \text{ or } \int 2x\,dx = x^2 - 5,$$

as well as
$$\int 2x\,dx = x^2;$$

since $x^2 + 2$ and $x^2 - 5$ are functions, each of whose differentials is $2x\,dx$.

In general
$$\int 2x\,dx = x^2 + c,$$

where c denotes an arbitrary constant called the *constant of integration*.

Every integral in its most general form includes this term, $+c$. We shall omit this constant of integration in the following integrals, as it can readily be added when necessary.

(*b*). Since $\qquad d(u \pm v \pm w) = du \pm dv \pm dw,$

it follows that

$$\int (du \pm dv \pm dw) = \int du \pm \int dv \pm \int dw.$$

That is, we integrate a polynomial by integrating the separate terms, and retaining the signs.

(*c*). Since $\qquad d(au) = a\,du,$

it follows that $\qquad \int a\,du = a \int du.$

That is, a constant factor may be transferred from one side of the symbol $\int$ to the other, without affecting the integral.

ELEMENTARY INTEGRALS. 175

3. *Fundamental Integrals.* We shall now give a list of formulæ, which may be regarded as fundamental, and to which all integrals must ultimately be reduced. We shall then consider in this chapter such examples as are integrable by these formulæ, either directly, or after some simple transformation.

I. $\int u^n du = \dfrac{u^{n+1}}{n+1}.$

II. $\int \dfrac{du}{u} = \log u.$

III. $\int a^u du = \dfrac{a^u}{\log a}.$

IV. $\int e^u du = e^u.$

V. $\int \cos u\, du = \sin u.$

VI. $\int \sin u\, du = -\cos u.$

VII. $\int \sec^2 u\, du = \tan u.$

VIII. $\int \operatorname{cosec}^2 u\, du = -\cot u.$

IX. $\int \sec u \tan u\, du = \sec u.$

X. $\int \operatorname{cosec} u \cot u\, du = -\operatorname{cosec} u.$

XI. $\int \tan u\, du = \log \sec u.$

XII. $\int \cot u\, du = \log \sin u.$

XIII. $\int \sec u\, du = \log(\sec u + \tan u) = \log \tan\left(\dfrac{\pi}{4} + \dfrac{u}{2}\right).$

XIV. $\int \operatorname{cosec} u\, du = \log(\operatorname{cosec} u - \cot u) = \log \tan \dfrac{u}{2}.$

XV. $\int \dfrac{du}{u^2 + a^2} = \dfrac{1}{a}\tan^{-1}\dfrac{u}{a}$, or $= -\dfrac{1}{a}\cot^{-1}\dfrac{u}{a}$.

XVI. $\int \dfrac{du}{u^2 - a^2} = \dfrac{1}{2a}\log\dfrac{u-a}{u+a}$, or $= \dfrac{1}{2a}\log\dfrac{a-u}{a+u}$.

XVII. $\int \dfrac{du}{\sqrt{a^2 - u^2}} = \sin^{-1}\dfrac{u}{a}$, or $= -\cos^{-1}\dfrac{u}{a}$.

XVIII. $\int \dfrac{du}{\sqrt{u^2 \pm a^2}} = \log(u + \sqrt{u^2 \pm a^2})$.

XIX. $\int \dfrac{du}{u\sqrt{u^2 - a^2}} = \dfrac{1}{a}\sec^{-1}\dfrac{u}{a}$, or $= -\dfrac{1}{a}\operatorname{cosec}^{-1}\dfrac{u}{a}$.

XX. $\int \dfrac{du}{\sqrt{2au - u^2}} = \operatorname{vers}^{-1}\dfrac{u}{a}$.

4. *Proof of* I. *and* II.

To derive I.,

since $\qquad d(u^{n+1}) = (n+1)u^n du,$

therefore

$$u^{n+1} = \int (n+1)u^n du = (n+1)\int u^n du, \qquad \text{by (c) Art. 2.}$$

Hence $\quad \int u^n du = \dfrac{u^{n+1}}{n+1}.$

Formula II. follows directly from

$$d\log u = \dfrac{du}{u}.$$

It is to be noticed that I. applies to all values of n except $n = -1$. For this value, it gives

$$\int u^{-1} du = \dfrac{u^0}{0} = \infty.$$

Formula II. provides for this failing case of I.

ELEMENTARY INTEGRALS.

EXAMPLES
For Formulæ I. and II.

Integrate the following expressions:

1. $\int x^4 dx = \frac{x^5}{5}$, by I., where $u = x$, and $n = 4$.

2. $\int (x^2 + 1)^{\frac{1}{2}} x\, dx$.

   If we apply I., calling $u = x^2 + 1$, and $n = \frac{1}{2}$; then $du = 2x\, dx$.

   We must then introduce a factor 2 before the $x\, dx$, and consequently its reciprocal $\frac{1}{2}$ on the left of $\int$.

   $$\int (x^2 + 1)^{\frac{1}{2}} x\, dx = \frac{1}{2} \int (x^2 + 1)^{\frac{1}{2}} 2x\, dx, \qquad \text{by (c) Art. 2.}$$
   $$= \frac{1}{2} \frac{(x^2 + 1)^{\frac{3}{2}}}{\frac{3}{2}} = \frac{(x^2 + 1)^{\frac{3}{2}}}{3}.$$

3. $\int \frac{(x^2 - a^2)\, dx}{x^3 - 3a^2 x} = \frac{1}{3} \int \frac{(3x^2 - 3a^2)\, dx}{x^3 - 3a^2 x}$
   $$= \frac{1}{3} \log(x^3 - 3a^2 x) = \log(x^3 - 3a^2 x)^{\frac{1}{3}}.$$

   By introducing the factor 3, we make the numerator the differential of the denominator, and then apply II.

4. $\int (x^5 + 3x^8 - 6x^{11})\, dx = \frac{1}{6}(x^6 + 2x^9 - 3x^{12})$.

5. $\int \left(10 x^{\frac{2}{3}} - \frac{1}{x^4}\right) dx = 6 x^{\frac{5}{3}} + \frac{1}{3x^3}$.

6. $\int \left(\frac{1}{\sqrt{x}} + \frac{1}{x^{\frac{3}{2}}}\right) dx = 2\left(\sqrt{x} - \frac{1}{\sqrt{x}}\right)$.

7. $\int \frac{x^2 - 1}{x}\, dx = \frac{x^2}{2} - \log x$.

8. $\int \frac{x\, dx}{x^2 - 1} = \log \sqrt{x^2 - 1}$.

9. $\int \dfrac{x+1}{x^2+2x} dx = \log \sqrt{x^2+2x}$.

10. $\int \dfrac{(x^2-2)^3 dx}{x^5} = \dfrac{2}{x^4} - \dfrac{6}{x^2} + \dfrac{x^2}{2} - \log x^6$.

11. $\int (a^2-x^2)^3 \sqrt{x}\, dx = 2x^{\frac{3}{2}}\left(\dfrac{a^6}{3} - \dfrac{3a^4 x^2}{7} + \dfrac{3a^2 x^4}{11} - \dfrac{x^6}{15}\right)$.

12. $\int (\sqrt{a} - \sqrt{x})^3 dx = a^{\frac{3}{2}} x - 2ax^{\frac{3}{2}} + \dfrac{3 a^{\frac{1}{2}} x^2}{2} - \dfrac{2 x^{\frac{5}{2}}}{5}$.

13. $\int (x+1)^2 dx = \dfrac{(x+1)^3}{3}$.

Integrate also, after expanding $(x+1)^2$. How are the two results reconciled?

14. $\int \dfrac{(x^n - a^n)^2 dx}{x} = \dfrac{x^n}{2n}(x^n - 4a^n) + a^{2n}\log x$.

15. $\int (x^2 - 2x + 2)(x-1) dx = \dfrac{(x^2 - 2x + 2)^2}{4}$.

Integrate also, after multiplying $x^2 - 2x + 2$ by $x-1$, and compare the two results.

16. $\int (3ax^2 - x^3)^{\frac{1}{2}} (2ax - x^2) dx = \dfrac{2}{15}(3ax^2 - x^3)^{\frac{5}{2}}$.

17. $\int \dfrac{(ax^2 + b) dx}{ax^3 + 3bx} = \dfrac{1}{3}\log(ax^3 + 3bx)$.

Integrate also, after multiplying numerator and denominator by 2, and compare the two results.

18. $\int \dfrac{dx}{(nx)^{\frac{n-1}{n}}} = (nx)^{\frac{1}{n}}$.

19. $\int \dfrac{x^{n-1} - 1}{x^n - nx} dx = \dfrac{1}{n}\log(x^n - nx)$.

20. $\int \left(\dfrac{x^2}{\sqrt{a^3+x^3}} - \dfrac{x}{\sqrt[4]{a^2+x^2}}\right) dx = \dfrac{2}{3}[(a^3+x^3)^{\frac{1}{2}} - (a^2+x^2)^{\frac{3}{4}}]$.

ELEMENTARY INTEGRALS.   179

21. $\displaystyle\int\frac{2x-1}{2x+3}dx = x - \log(2x+3)^2.$

22. $\displaystyle\int\frac{x^3 dx}{x+1} = x - \frac{x^2}{2} + \frac{x^3}{3} - \log(x+1).$

23. $\displaystyle\int\frac{dx}{(\sqrt{a}+\sqrt{x})^{\frac{1}{2}}\sqrt{x}} = 4(\sqrt{a}+\sqrt{x})^{\frac{1}{2}}.$

24. $\displaystyle\int\frac{(a^{\frac{1}{3}}-x^{\frac{1}{3}})^{\frac{1}{2}}dx}{x^{\frac{2}{3}}} = -2(a^{\frac{1}{3}}-x^{\frac{1}{3}})^{\frac{3}{2}}.$

25. $\displaystyle\int\frac{\log(x+1)}{x+1}dx = \frac{1}{2}[\log(x+1)]^2.$

26. $\displaystyle\int\frac{dx}{(x+a)^{\frac{1}{2}}+(x+b)^{\frac{1}{2}}} = \frac{2}{3(a-b)}[(x+a)^{\frac{3}{2}} - (x+b)^{\frac{3}{2}}].$

27. $\displaystyle\int (x^3+1)(x^3+5)^{\frac{1}{3}}dx = \frac{x}{5}(x^3+5)^{\frac{4}{3}}.$

  Suggestion. $(x^3+1)(x^3+5)^{\frac{1}{3}} = (x^{\frac{15}{4}}+5x^{\frac{3}{4}})^{\frac{1}{3}}(x^{\frac{11}{4}}+x^{-\frac{1}{4}}).$

28. $\displaystyle\int\frac{(x^n+1)dx}{(x^n+n)^{\frac{1}{n}}} = \frac{x}{n}(x^n+n)^{\frac{n-1}{n}}.$

  Suggestion. Multiply numerator and denominator by $x^{\frac{1}{n-1}}$.

The following integrals may be evaluated by I., after multiplying the binomial under the radical sign by x^{-2}.

29. $\displaystyle\int\frac{dx}{x^2\sqrt{a^2-x^2}} = \int\frac{x^{-3}dx}{\sqrt{a^2x^{-2}-1}} = \int (a^2x^{-2}-1)^{-\frac{1}{2}} x^{-3}dx$

$\displaystyle = -\frac{1}{2a^2}\int (a^2x^{-2}-1)^{-\frac{1}{2}}(-2a^2x^{-3}dx)$

$\displaystyle = -\frac{1}{2a^2}\frac{(a^2x^{-2}-1)^{\frac{1}{2}}}{\frac{1}{2}} = -\frac{(a^2x^{-2}-1)^{\frac{1}{2}}}{a^2}$

$\displaystyle = -\frac{\sqrt{a^2-x^2}}{a^2x}.$

30. $\displaystyle\int \frac{dx}{x^2\sqrt{x^2+a^2}} = -\frac{\sqrt{x^2+a^2}}{a^2 x}.$

31. $\displaystyle\int \frac{\sqrt{a^2-x^2}\,dx}{x^4} = -\frac{(a^2-x^2)^{\frac{3}{2}}}{3\,a^2 x^3}.$

32. $\displaystyle\int \frac{\sqrt{x^2-a^2}\,dx}{x^4} = \frac{(x^2-a^2)^{\frac{3}{2}}}{3\,a^2 x^3}.$

33. $\displaystyle\int \frac{dx}{(a^2-x^2)^{\frac{3}{2}}} = \frac{x}{a^2\sqrt{a^2-x^2}}.$

34. $\displaystyle\int \frac{dx}{(x^2+a^2)^{\frac{3}{2}}} = \frac{x}{a^2\sqrt{x^2+a^2}}.$

35. $\displaystyle\int \frac{dx}{x\sqrt{2ax-x^2}} = -\frac{\sqrt{2ax-x^2}}{ax}.$

36. $\displaystyle\int \frac{x\,dx}{(2ax-x^2)^{\frac{3}{2}}} = \frac{x}{a\sqrt{2ax-x^2}}.$

37. $\displaystyle\int \frac{\sqrt{2ax-x^2}\,dx}{x^3} = -\frac{(2ax-x^2)^{\frac{3}{2}}}{3\,ax^3}.$

38. $\displaystyle\int \frac{dx}{(2ax-x^2)^{\frac{3}{2}}} = \frac{x-a}{a^2\sqrt{2ax-x^2}}.$

This may be obtained from Ex. 33 by substituting $x-a$ for x.

5. Proof of III. and IV. These are evidently obtained directly from the corresponding formulæ of differentiation.

EXAMPLES

For Formulæ III. and IV.

1. $\displaystyle\int (e^{3x} + a^{3x})\,dx = \frac{1}{3}\left(e^{3x} + \frac{a^{3x}}{\log a}\right).$

2. $\displaystyle\int (e^{ax} + e^{\frac{x}{a}})\,dx = \frac{e^{ax}}{a} + a e^{\frac{x}{a}}.$

3. $\displaystyle\int (a^{nx} - b^{mx})\,dx = \frac{a^{nx}}{n\log a} - \frac{b^{mx}}{m\log b}.$

4. $\int (e^x + e^{-x})^2 dx = \frac{1}{2}(e^{2x} - e^{-2x}) + 2x.$

5. $\int \frac{(e^x+1)^2}{\sqrt{e^x}} dx = \frac{2}{\sqrt{e^x}}\left(\frac{e^{2x}}{3} + 2e^x - 1\right).$

6. $\int (3e^{2x} - 1)^{\frac{1}{3}} e^{2x} dx = \frac{1}{8}(3e^{2x} - 1)^{\frac{4}{3}}.$

7. $\int \frac{e^{3x} dx}{e^x - 1} = \frac{e^{2x}}{2} + e^x + \log(e^x - 1).$

8. $\int \frac{e^x - 1}{e^x + 1} dx = \log(e^x + 1)^2 - x.$

9. $\int a^x e^x dx = \frac{a^x e^x}{1 + \log a}.$

10. $\int \frac{(a^x - b^x)^2}{a^x b^x} dx = \frac{a^x b^{-x} - a^{-x} b^x}{\log a - \log b} - 2x.$

6. *Proof of* V.–XIV. It is evident that V.–X. are obtained directly from the corresponding formulæ of differentiation.

To derive XI. and XII.,

$$\int \tan u \, du = -\int \frac{-\sin u \, du}{\cos u} = -\log \cos u = \log \sec u.$$

$$\int \cot u \, du = \int \frac{\cos u \, du}{\sin u} = \log \sin u.$$

To derive XIII. and XIV.,

$$\int \sec u \, du = \int \frac{\sec u (\tan u + \sec u) du}{\sec u + \tan u} = \int \frac{\sec u \tan u \, du + \sec^2 u \, du}{\sec u + \tan u}$$

$$= \log(\sec u + \tan u).$$

$$\int \cosec u \, du = \int \frac{\cosec u (-\cot u + \cosec u) du}{\cosec u - \cot u}$$

$$= \log(\cosec u - \cot u).$$

INTEGRAL CALCULUS.

By Trigonometry,

$$\operatorname{cosec} u - \cot u = \frac{1-\cos u}{\sin u} = \frac{2\sin^2 \frac{u}{2}}{2\sin \frac{u}{2}\cos \frac{u}{2}} = \tan \frac{u}{2}.$$

If we substitute in this $\frac{\pi}{2}+u$ for u,

we have $\sec u + \tan u = \tan\left(\frac{\pi}{4}+\frac{u}{2}\right).$

Hence we obtain the second forms of XIII. and XIV.

EXAMPLES
For Formulæ V.–XIV.

1. $\int (\sin 2x + \cos 2x)dx = \frac{1}{2}(\sin 2x - \cos 2x).$

2. $\int \left(\cos \frac{x}{3} - \sin 3x\right)dx = 3\sin \frac{x}{3} + \frac{1}{3}\cos 3x.$

3. $\int [\sin(a+bx) - \cos(a-bx)]dx = \frac{\sin(a-bx) - \cos(a+bx)}{b}.$

4. $\int \frac{\sin 3x\, dx}{\cos^2 3x} = \frac{1}{3}\sec 3x.$

5. $\int \sec \frac{x}{2}\left(\sec \frac{x}{2} + \tan \frac{x}{2}\right)dx = 2\left(\tan \frac{x}{2} + \sec \frac{x}{2}\right).$

6. $\int \frac{1-\cos ax}{\sin^2 ax}dx = \frac{1}{a}(\operatorname{cosec} ax - \cot ax).$

7. $\int (\tan x + \cot x)^2 dx = \tan x - \cot x.$

8. $\int (\sec x - \tan x)^2 dx = 2(\tan x - \sec x) - x.$

9. $\int \frac{\sin x\, dx}{a + b\cos x} = -\frac{1}{b}\log(a + b\cos x).$

10. $\int \frac{\tan x\, dx}{a + b\tan^2 x} = \frac{\log(a\cos^2 x + b\sin^2 x)}{2(b-a)}.$

11. $\int (\tan 2x - 1)^2 dx = \frac{1}{2}\tan 2x + \log \cos 2x.$

12. $\int (\sec 2x + 1)^2 dx = \frac{1}{2}\tan 2x + \log(\sec 2x + \tan 2x) + x.$

13. $\int (\cosec x - 1)(\cot x + 1) dx = -x - \cosec x - \log(1 + \cos x).$

14. $\int (\sec x + \cosec x)^2 dx = \tan x - \cot x + 2\log \tan x.$

15. $\int \sin^2 x\, dx = \frac{x}{2} - \frac{1}{4}\sin 2x.$

16. $\int \cos^2 x\, dx = \frac{x}{2} + \frac{1}{4}\sin 2x.$

17. $\int \frac{1 + \sin x}{1 - \sin x} dx = 2(\sec x + \tan x) - x.$

18. $\int \frac{\cot x + \tan x}{\cot x - \tan x} dx = \frac{1}{2}\log \tan\left(\frac{\pi}{4} + x\right).$

19. $\int \tan x \tan(x + a)\, dx = -x - \frac{\log(1 - \tan a \tan x)}{\tan a}.$

20. $\int \sec x \sec(x + a)\, dx = \frac{1}{\sin a}\log \frac{\cos x}{\cos(x + a)}.$

7. Proof of XV.–XX.

To derive XV.,

$$\int \frac{du}{u^2 + a^2} = \frac{1}{a}\int \frac{\frac{du}{a}}{1 + \frac{u^2}{a^2}} = \frac{1}{a}\int \frac{d\left(\frac{u}{a}\right)}{1 + \left(\frac{u}{a}\right)^2} = \frac{1}{a}\tan^{-1}\frac{u}{a}.$$

To derive XVII.,

$$\int \frac{du}{\sqrt{a^2 - u^2}} = \int \frac{\frac{du}{a}}{\sqrt{1 - \frac{u^2}{a^2}}} = \sin^{-1}\frac{u}{a}.$$

INTEGRAL CALCULUS.

To derive XIX.,

$$\int \frac{du}{u\sqrt{u^2-a^2}} = \frac{1}{a}\int \frac{\frac{du}{a}}{\frac{u}{a}\sqrt{\frac{u^2}{a^2}-1}} = \frac{1}{a}\sec^{-1}\frac{u}{a}.$$

To derive XX.,

$$\int \frac{du}{\sqrt{2au-u^2}} = \int \frac{\frac{du}{a}}{\sqrt{2\frac{u}{a}-\frac{u^2}{a^2}}} = \text{vers}^{-1}\frac{u}{a}.$$

Since $\tan^{-1}\frac{u}{a} = \frac{\pi}{2} - \cot^{-1}\frac{u}{a},$

it is evident that $d\tan^{-1}\frac{u}{a} = d\left(-\cot^{-1}\frac{u}{a}\right).$

Hence either expression may be used as the integral in XV.

In the same way we obtain the second forms of XVII. and XIX.

The formulæ XVI. and XVIII. are inserted in the list of integrals, because they are of similar form to XV. and XVII., respectively, with different signs.

To derive XVI.,

$$\frac{1}{u^2-a^2} = \frac{1}{2a}\left(\frac{1}{u-a} - \frac{1}{u+a}\right);$$

hence

$$\int \frac{du}{u^2-a^2} = \frac{1}{2a}\int\left(\frac{du}{u-a} - \frac{du}{u+a}\right)$$
$$= \frac{1}{2a}[\log(u-a) - \log(u+a)] = \frac{1}{2a}\log\frac{u-a}{u+a}.$$

Or we may integrate thus:

$$\int \frac{du}{u^2-a^2} = \frac{1}{2a}\int\left(\frac{-du}{a-u} - \frac{du}{a+u}\right)$$
$$= \frac{1}{2a}[\log(a-u) - \log(a+u)] = \frac{1}{2a}\log\frac{a-u}{a+u}.$$

ELEMENTARY INTEGRALS.

To derive XVIII., assume $\sqrt{u^2 \pm a^2} = z$, a new variable.

Then
$$u^2 \pm a^2 = z^2,$$
$$2u\,du = 2z\,dz;$$

therefore
$$\frac{du}{z} = \frac{dz}{u} = \frac{du+dz}{u+z}.$$

Hence
$$\int \frac{du}{z} = \int \frac{du+dz}{u+z} = \log(u+z);$$

that is,
$$\int \frac{du}{\sqrt{u^2 \pm a^2}} = \log(u + \sqrt{u^2 \pm a^2}).$$

EXAMPLES
For Formulæ XV.–XX.

1. $\int \dfrac{dx}{9x^2+4} = \dfrac{1}{6}\tan^{-1}\dfrac{3x}{2}.$

2. $\int \dfrac{dx}{9x^2-4} = \dfrac{1}{12}\log\dfrac{3x-2}{3x+2}.$

3. $\int \dfrac{dx}{\sqrt{1-4x^2}} = \dfrac{1}{2}\sin^{-1}2x.$

4. $\int \dfrac{dx}{\sqrt{1+4x^2}} = \dfrac{1}{2}\log(2x + \sqrt{1+4x^2}).$

5. $\int \dfrac{x\,dx}{\sqrt{1-x^4}} = \dfrac{1}{2}\sin^{-1}x^2.$

6. $\int \dfrac{x\,dx}{x^4+4} = \dfrac{1}{4}\tan^{-1}\dfrac{x^2}{2}.$

7. $\int \dfrac{x\,dx}{x^4-4} = \dfrac{1}{8}\log\dfrac{x^2-2}{x^2+2}.$

8. $\int \dfrac{dx}{x\sqrt{4x^2-9}} = \dfrac{1}{3}\sec^{-1}\dfrac{2x}{3}.$

9. $\int \dfrac{dx}{\sqrt{6x-x^2}} = \text{vers}^{-1}\dfrac{x}{3}.$

10. $\int \dfrac{dx}{\sqrt{ax-b^2x^2}} = \dfrac{1}{b}\operatorname{vers}^{-1}\dfrac{2b^2x}{a}$.

11. $\int \dfrac{dx}{x\sqrt{a^2x^2-b^2}} = \dfrac{1}{b}\sec^{-1}\dfrac{ax}{b}$.

12. $\int \dfrac{dx}{\sqrt{2x-3x^2}} = \dfrac{1}{\sqrt{3}}\operatorname{vers}^{-1}3x$.

13. $\int \dfrac{dx}{a^2-b^2x^2} = -\dfrac{1}{2ab}\log\dfrac{bx-a}{bx+a} = \dfrac{1}{2ab}\log\dfrac{bx+a}{bx-a}$.

14. $\int \dfrac{2x-5}{3x^2+2}dx = \dfrac{1}{3}\log(3x^2+2) - \dfrac{5}{\sqrt{6}}\tan^{-1}\dfrac{3x}{\sqrt{6}}$.

15. $\int \dfrac{2x-5}{3x^2-2}dx = \dfrac{1}{3}\log(3x^2-2) - \dfrac{5}{2\sqrt{6}}\log\dfrac{x\sqrt{3}-\sqrt{2}}{x\sqrt{3}+\sqrt{2}}$.

The same formulæ may be applied to expressions involving x^2+ax+b or $-x^2+ax+b$, by completing the square with the terms containing x. Thus, —

16. $\int \dfrac{dx}{x^2+2x+5} = \int \dfrac{dx}{(x+1)^2+4} = \dfrac{1}{2}\tan^{-1}\dfrac{x+1}{2}$.

17. $\int \dfrac{dx}{\sqrt{2+x-x^2}} = \int \dfrac{2\,dx}{\sqrt{8+4x-4x^2}} = \int \dfrac{2\,dx}{\sqrt{9-(2x-1)^2}}$
$= \sin^{-1}\dfrac{2x-1}{3}$.

18. $\int \dfrac{dx}{x^2-6x+11} = \dfrac{1}{\sqrt{2}}\tan^{-1}\dfrac{x-3}{\sqrt{2}}$.

19. $\int \dfrac{dx}{x^2-6x+5} = \dfrac{1}{4}\log\dfrac{x-5}{x-1}$.

20. $\int \dfrac{dx}{x^2+3x+1} = \dfrac{1}{\sqrt{5}}\log\dfrac{2x+3-\sqrt{5}}{2x+3+\sqrt{5}}$.

21. $\int \dfrac{dx}{5x^2-2x+1} = \dfrac{1}{2}\tan^{-1}\dfrac{5x-1}{2}$.

22. $\int \dfrac{dx}{\sqrt{1+3x-x^2}} = \sin^{-1}\dfrac{2x-3}{\sqrt{13}}.$

23. $\int \dfrac{dx}{\sqrt{x^2-4x+13}} = \log(x-2+\sqrt{x^2-4x+13}).$

24. $\int \dfrac{dx}{x^2-2x\sin a+1} = \sec a \tan^{-1}(x\sec a-\tan a).$

25. $\int \dfrac{dx}{\sqrt{3x^2-4x}} = \dfrac{1}{\sqrt{3}}\log(3x-2+\sqrt{9x^2-12x}).$

26. $\int \dfrac{2\,dx}{3x^2+10x+3} = \dfrac{1}{4}\log\dfrac{3x+1}{x+3}.$

27. $\int \dfrac{dx}{ax^2+bx+c} = \dfrac{2}{\sqrt{4ac-b^2}}\tan^{-1}\dfrac{2ax+b}{\sqrt{4ac-b^2}},$

    or $= \dfrac{1}{\sqrt{b^2-4ac}}\log\dfrac{2ax+b-\sqrt{b^2-4ac}}{2ax+b+\sqrt{b^2-4ac}}.$

28. $\int \dfrac{dx}{\sqrt{ax^2+bx+c}} = \dfrac{1}{\sqrt{a}}\log(2ax+b+2\sqrt{a}\sqrt{ax^2+bx+c}).$

29. $\int \dfrac{dx}{\sqrt{-ax^2+bx+c}} = \dfrac{1}{\sqrt{a}}\sin^{-1}\dfrac{2ax-b}{\sqrt{b^2+4ac}}.$

CHAPTER II.

INTEGRATION OF RATIONAL FRACTIONS.

8. *Preliminary Operation.* If the degree of the numerator is equal to, or greater than, that of the denominator, the fraction should be reduced to a mixed quantity, by dividing the numerator by the denominator.

For example,

$$\frac{x^3 - 2x^2}{x^3 + 1} = 1 - \frac{2x^2 + 1}{x^3 + 1},$$

$$\frac{2x^5 - 3x^4 + 1}{x^4 + x^2} = 2x - 3 + \frac{-2x^3 + 3x^2 + 1}{x^4 + x^2}.$$

The degree of the numerator of this new fraction will be less than that of the denominator. Such fractions only will be considered in the following articles.

9. *Factors of the Denominator.* A rational fraction is integrated by decomposing it into partial fractions, whose denominators are the factors of the original denominator.

Now it is shown by the Theory of Equations, that a polynomial of the nth degree with respect to x, may be resolved into n factors of the first degree,

$$(x - a_1)(x - a_2)(x - a_3) \cdots (x - a_n).$$

These factors are real or imaginary, but the imaginary factors will occur in pairs, of the form

$$x - a + b\sqrt{-1}, \text{ and } x - a - b\sqrt{-1},$$

whose product is $(x - a)^2 + b^2$, a real factor of the second degree.

It follows, then, that any polynomial may be resolved into real factors of the first or second degree, and only such factors will be considered in the denominators of fractions.

There are four cases to be considered.

First. Where the denominator contains factors of the *first* degree only, each of which occurs but once.

Second. Where the denominator contains factors of the *first* degree only, some of which are repeated.

Third. Where the denominator contains factors of the *second* degree, each of which occurs but once.

Fourth. Where the denominator contains factors of the *second* degree, some of which are repeated.

10. CASE I. *Factors of the denominator all of the first degree, and none repeated.*

The given fraction may be decomposed into partial fractions, as shown by the following example,

Assume $$\int \frac{x^2 + 6x - 8}{x^3 - 4x} dx.$$

$$\frac{x^2 + 6x - 8}{x^3 - 4x} = \frac{x^2 + 6x - 8}{x(x-2)(x+2)} = \frac{A}{x} + \frac{B}{x-2} + \frac{C}{x+2}, \quad (1)$$

where A, B, C, are unknown constants.

Clearing (1) of fractions,

$$x^2 + 6x - 8 = A(x-2)(x+2) + Bx(x+2) + Cx(x-2) \quad (2)$$
$$= (A + B + C)x^2 + 2(B - C)x - 4A.$$

Equating the coefficients of like powers of x in the two members of the equation, according to the method of Indeterminate Coefficients, we have

$$A + B + C = 1,$$
$$2(B - C) = 6,$$
$$-4A = -8;$$

whence $\qquad A = 2, \quad B = 1, \quad C = -2.$

Hence $\dfrac{x^2+6x-8}{x^3-4x} = \dfrac{2}{x} + \dfrac{1}{x-2} - \dfrac{2}{x+2}$;

and $\displaystyle\int \dfrac{x^2+6x-8}{x^3-4x} dx = 2\log x + \log(x-2) - 2\log(x+2)$

$$= \log\dfrac{x^2(x-2)}{(x+2)^2}.$$

A shorter method of finding A, B, C, is the following:

If in (2) we let $x=0$, B and C will disappear from the equation, and we shall have

$$-8 = -4A, \text{ or } A = 2.$$

Similarly, If $x=2$, $\quad 8 = 8B,\quad$ or $B=1$.

If $x=-2$, $-16 = 8C$, or $C=-2$.

EXAMPLES.

1. $\displaystyle\int \dfrac{3x-1}{x^2+x-6} dx = \log[(x+3)^2(x-2)]$.

2. $\displaystyle\int \dfrac{1+x^2}{x-x^3} dx = \log\dfrac{x}{1-x^2}$.

3. $\displaystyle\int \dfrac{x^2+2x-\cos^2 a}{x^2+2x+\sin^2 a} dx = x + \dfrac{\sec a}{2}\log\dfrac{x+1+\cos a}{x+1-\cos a}$.

4. $\displaystyle\int \dfrac{x^4 dx}{(x^2-1)(x+2)} = \dfrac{x^2}{2} - 2x + \dfrac{1}{6}\log\dfrac{x-1}{(x+1)^3} + \dfrac{16}{3}\log(x+2)$.

5. $\displaystyle\int \dfrac{x\,dx}{x^2-4x+1}$

$= \dfrac{2+\sqrt{3}}{2\sqrt{3}}\log(x-2-\sqrt{3}) - \dfrac{2-\sqrt{3}}{2\sqrt{3}}\log(x-2+\sqrt{3})$

$= \dfrac{1}{2}\log(x^2-4x+1) + \dfrac{1}{\sqrt{3}}\log\dfrac{x-2-\sqrt{3}}{x-2+\sqrt{3}}$.

6. $\int \frac{x^3 + x^4 - 8}{x^3 - 4x} dx = \frac{x^3}{3} + \frac{x^2}{2} + 4x + \log \frac{x^2(x-2)^5}{(x+2)^3}.$

7. $\int \frac{6(x+3)dx}{x^5 - 5x^3 + 4x} = \log\left[\frac{x^{\frac{3}{2}}(x-2)^{\frac{5}{4}}(x+2)^{\frac{1}{4}}}{(x-1)^4(x+1)^2}\right].$

11. Case II. *Factors of the denominator all of the first degree, and some repeated.*

Here the method of decomposition of Case I. requires modification. Suppose, for example, we have

$$\int \frac{x^3 + 1}{x(x-1)^3} dx.$$

If we follow the method of the preceding case, we should write
$$\frac{x^3 + 1}{x(x-1)^3} = \frac{A}{x} + \frac{B}{x-1} + \frac{C}{x-1} + \frac{D}{x-1}.$$

But since the common denominator of the fractions in the second member of this equation is $x(x-1)$, their sum cannot be equal to the given fraction with the denominator $x(x-1)^3$. To meet this objection, we assume

$$\frac{x^3 + 1}{x(x-1)^3} = \frac{A}{x} + \frac{B}{(x-1)^3} + \frac{C}{(x-1)^2} + \frac{D}{x-1}.$$

Clearing of fractions,
$$x^3 + 1 = A(x-1)^3 + Bx + Cx(x-1) + Dx(x-1)^2$$
$$= (A+D)x^3 + (-3A + C - 2D)x^2$$
$$+ (3A + B - C + D)x - A.$$

Hence
$$A + D = 1,$$
$$-3A + C - 2D = 0,$$
$$3A + B - C + D = 0,$$
$$-A = 1.$$

Whence $A = -1,\ B = 2,\ C = 1,\ D = 2.$

Therefore $\dfrac{x^3+1}{x(x-1)^3} = -\dfrac{1}{x} + \dfrac{2}{(x-1)^3} + \dfrac{1}{(x-1)^2} + \dfrac{2}{x-1}$.

Hence

$$\int \dfrac{x^3+1}{x(x-1)^3}dx = -\log x - \dfrac{1}{(x-1)^2} - \dfrac{1}{x-1} + 2\log(x-1)$$

$$= -\dfrac{x}{(x-1)^2} + \log\dfrac{(x-1)^2}{x}.$$

EXAMPLES.

1. $\displaystyle\int \dfrac{(x-8)dx}{x^3 - 4x^2 + 4x} = \dfrac{3}{x-2} + \log\dfrac{(x-2)^2}{x^2}$.

2. $\displaystyle\int \dfrac{3x^2 - 2}{(x+2)^3}dx = \dfrac{12x + 19}{(x+2)^2} + 3\log(x+2)$.

3. $\displaystyle\int \dfrac{(3x+2)dx}{x(x+1)^3} = \dfrac{4x+3}{2(x+1)^2} + \log\dfrac{x^2}{(x+1)^2}$.

4. $\displaystyle\int \dfrac{x^5 - 5x - 3}{(x^2+x)^2}dx = \dfrac{x^2}{2} - 2x + \dfrac{2x+3}{x^2+x} + \log[x(x+1)^2]$.

5. $\displaystyle\int \dfrac{dx}{(x^2-2)^2} = -\dfrac{x}{4(x^2-2)} + \dfrac{1}{8\sqrt{2}}\log\dfrac{x+\sqrt{2}}{x-\sqrt{2}}$.

6. $\displaystyle\int \dfrac{9(-x^2+4x+2)\,dx}{(x^3-x-2)^3} = \dfrac{2x-5}{(x-2)^2} + \dfrac{2x+1}{2(x+1)^2} + \log\dfrac{x-2}{x+1}$.

7. $\displaystyle\int \dfrac{(8x^5-1)\,dx}{(2x^2-x)^3} = x - \dfrac{12x+1}{2x^2} - \dfrac{108x - 61}{4(2x-1)^2} + 24\log x$

$\qquad\qquad - \dfrac{45}{2}\log(2x-1)$.

12. Case III. *Denominator containing factors of the second degree, but none repeated.*

The form of decomposition will appear from the following example,

$$\int \dfrac{5x+12}{x(x^2+4)}dx.$$

We assume
$$\frac{5x+12}{x(x^2+4)} = \frac{A}{x} + \frac{Bx+C}{x^2+4}, \quad \cdots \cdots \quad (1)$$

and in general for every partial fraction in this case, whose denominator is of the second degree, we must assume a numerrator of the form $Bx+C$.

For it is evident that each additional fraction of this kind increases by two the degree of the equation, when cleared of fractions, and consequently increases by two the number of equations for determining $A, B, C, \cdots$.

Hence its numerator should add two to the number of these unknown quantities.

Clearing (1) of fractions,

$$5x+12 = (A+B)x^2 + Cx + 4A.$$

$$A+B=0, \quad C=5, \quad 4A=12.$$

Whence $A=3, \ B=-3, \ C=5$;

therefore
$$\frac{5x+12}{x(x^2+4)} = \frac{3}{x} + \frac{-3x+5}{x^2+4}.$$

$$\int \frac{-3x+5}{x^2+4} dx = -3 \int \frac{x\,dx}{x^2+4} + 5 \int \frac{dx}{x^2+4}$$

$$= -\frac{3}{2}\log(x^2+4) + \frac{5}{2}\tan^{-1}\frac{x}{2}.$$

Hence
$$\int \frac{5x+12}{x(x^2+4)} dx = 3\log\frac{x}{\sqrt{x^2+4}} + \frac{5}{2}\tan^{-1}\frac{x}{2}.$$

Take for another example,

$$\int \frac{(2x^2-3x-3)dx}{(x-1)(x^2-2x+5)}.$$

This fraction is decomposed as follows:

$$\frac{2x^2-3x-3}{(x-1)(x^2-2x+5)} = -\frac{1}{x-1} + \frac{3x-2}{x^2-2x+5}.$$

$$\int \frac{(3x-2)\,dx}{x^2-2x+5} = \int \frac{(3x-3)\,dx}{x^2-2x+5} + \int \frac{dx}{x^2-2x+5}$$

$$= \frac{3}{2}\log(x^2-2x+5) + \frac{1}{2}\tan^{-1}\frac{x-1}{2}.$$

$$\int \frac{(2x^2-3x-3)\,dx}{(x-1)(x^2-2x+5)} = \log\frac{(x^2-2x+5)^{\frac{3}{2}}}{x-1} + \frac{1}{2}\tan^{-1}\frac{x-1}{2}.$$

EXAMPLES.

1. $\int \frac{x^3-1}{x^3+3x}\,dx = x + \frac{1}{6}\log\frac{x^2+3}{x^2} - \sqrt{3}\tan^{-1}\frac{x}{\sqrt{3}}.$

2. $\int \frac{dx}{(x^2+1)(x^2+x)} = \frac{1}{4}\log\frac{x^4}{(x+1)^2(x^2+1)} - \frac{1}{2}\tan^{-1}x.$

3. $\int \frac{x^2\,dx}{(x-1)^2(x^2+1)} = -\frac{1}{2(x-1)} + \frac{1}{4}\log\frac{(x-1)^2}{x^2+1}.$

4. $\int \frac{(x^3-6)\,dx}{x^4+6x^2+8} = \log\frac{x^2+4}{\sqrt{x^2+2}} + \frac{3}{2}\tan^{-1}\frac{x}{2} - \frac{3}{\sqrt{2}}\tan^{-1}\frac{x}{\sqrt{2}}.$

5. $\int \frac{dx}{x^4-1} = \frac{1}{4}\log\frac{x-1}{x+1} - \frac{1}{2}\tan^{-1}x.$

6. $\int \frac{(5x^2-1)\,dx}{(x^2+3)(x^2-2x+5)}$

$$= \log\frac{x^2-2x+5}{x^2+3} + \frac{5}{2}\tan^{-1}\frac{x-1}{2} - \frac{2}{\sqrt{3}}\tan^{-1}\frac{x}{\sqrt{3}}.$$

7. $\int \frac{(9x-10)\,dx}{x^2(2x^2-2x+5)} = \frac{2}{x} + \frac{1}{2}\log\frac{x^3}{2x^2-2x+5} + \frac{5}{3}\tan^{-1}\frac{2x-1}{3}.$

8. $\int \frac{dx}{x^3+1} = \frac{1}{6}\log\frac{(x+1)^2}{x^2-x+1} + \frac{1}{\sqrt{3}}\tan^{-1}\frac{2x-1}{\sqrt{3}}.$

9. $\int \frac{(3x^2-5)\,dx}{x^4+6x^2+25}$

$$= \frac{1}{2}\log\frac{x^2-2x+5}{x^2+2x+5} + \frac{1}{4}\left(\tan^{-1}\frac{x+1}{2} + \tan^{-1}\frac{x-1}{2}\right)$$

$$= \frac{1}{2}\log\frac{x^2-2x+5}{x^2+2x+5} + \frac{1}{4}\tan^{-1}\frac{4x}{5-x^2}.$$

RATIONAL FRACTIONS. 195

10. $\int \dfrac{4\,dx}{x^4+1} = \dfrac{1}{\sqrt{2}} \log \dfrac{x^2+x\sqrt{2}+1}{x^2-x\sqrt{2}+1} + \sqrt{2}\tan^{-1}\dfrac{x\sqrt{2}}{1-x^2}$.

11. $\int \dfrac{x^2\cos 2a+1}{x^4+2x^2\cos 2a+1}\,dx$

$= \dfrac{\sin a}{4}\log\dfrac{x^2+2x\sin a+1}{x^2-2x\sin a+1} + \dfrac{\cos a}{2}\tan^{-1}\dfrac{2x\cos a}{1-x^2}$.

13. CASE IV. *Denominator containing factors of the second degree, some of which are repeated.*

This case bears the same relation to Case III., that Case II. bears to Case I., and requires a similar modification of the partial fractions.

For illustration take
$$\int \dfrac{2x^3+x+3}{(x^2+1)^2}\,dx.$$

We assume
$$\dfrac{2x^3+x+3}{(x^2+1)^2} = \dfrac{Ax+B}{(x^2+1)^2} + \dfrac{Cx+D}{x^2+1}.$$

$2x^3+x+3 = Cx^3+Dx^2+(A+C)x+B+D$.

$A=-1,\quad B=3,\quad C=2,\quad D=0$.

Therefore $\dfrac{2x^3+x+3}{(x^2+1)^2} = \dfrac{-x+3}{(x^2+1)^2} + \dfrac{2x}{x^2+1}$.

$\int\dfrac{-x+3}{(x^2+1)^2}\,dx = -\int\dfrac{x\,dx}{(x^2+1)^2} + 3\int\dfrac{dx}{(x^2+1)^2}$

$\qquad = \dfrac{1}{2(x^2+1)} + 3\int\dfrac{dx}{(x^2+1)^2}$.

To integrate the last fraction, we use the following formula of reduction,

$\int\dfrac{dx}{(x^2+a^2)^n} = \dfrac{1}{2(n-1)a^2}\left[\dfrac{x}{(x^2+a^2)^{n-1}} + (2n-3)\int\dfrac{dx}{(x^2+a^2)^{n-1}}\right]$.

This formula will be derived in Chapter IV., but the student can now verify it by differentiating both members. It enables us to integrate the expression peculiar to this case, $\int \frac{dx}{(x^2+a^2)^n}$, by making it depend upon $\int \frac{dx}{(x^2+a^2)^{n-1}}$. By successive applications the given integral is made to depend ultimately upon $\int \frac{dx}{x^2+a^2}$, which is $\frac{1}{a}\tan^{-1}\frac{x}{a}$.

To apply this formula to $\int \frac{dx}{(x^2+1)^2}$, we make $a=1$ and $n=2$.

We then have

$$\int \frac{dx}{(x^2+1)^2} = \frac{1}{2}\left[\frac{x}{x^2+1} + \int \frac{dx}{x^2+1}\right] = \frac{x}{2(x^2+1)} + \frac{1}{2}\tan^{-1}x;$$

whence
$$\int \frac{-x+3}{(x^2+1)^2}dx = \frac{1}{2(x^2+1)} + \frac{3x}{2(x^2+1)} + \frac{3}{2}\tan^{-1}x,$$

and
$$\int \frac{2x^3+x+3}{(x^2+1)^2}dx = \frac{3x+1}{2(x^2+1)} + \frac{3}{2}\tan^{-1}x + \log(x^2+1).$$

As another example in the integration of a partial fraction in Case IV., consider

$$\int \frac{3x+2}{(x^2-3x+3)^2}dx = \int \frac{\left(3x-\frac{9}{2}\right)dx}{(x^2-3x+3)^2} + \frac{13}{2}\int \frac{dx}{(x^2-3x+3)^2}.$$

$$\int \frac{\left(3x-\frac{9}{2}\right)dx}{(x^2-3x+3)^2} = \frac{3}{2}\int \frac{(2x-3)dx}{(x^2-3x+3)^2} = -\frac{3}{2(x^2-3x+3)}.$$

$$\int \frac{dx}{(x^2-3x+3)^2} = \int \frac{dx}{\left[\left(x-\frac{3}{2}\right)^2+\frac{3}{4}\right]^2} = \int \frac{dz}{\left(z^2+\frac{3}{4}\right)^2},$$

where $z = x - \frac{3}{2}$.

Applying the formula of reduction,

$$\int \frac{dz}{\left(z^2+\frac{3}{4}\right)^2} = \frac{2}{3}\left(\frac{z}{z^2+\frac{3}{4}} + \int \frac{dz}{z^2+\frac{3}{4}}\right) = \frac{2}{3}\frac{z}{z^2+\frac{3}{4}} + \frac{4}{3\sqrt{3}}\tan^{-1}\frac{2z}{\sqrt{3}}.$$

Or substituting $z = x - \frac{3}{2}$,

$$\int \frac{dx}{(x^2-3x+3)^2} = \frac{2x-3}{3(x^2-3x+3)} + \frac{4}{3\sqrt{3}}\tan^{-1}\frac{2x-3}{\sqrt{3}};$$

hence

$$\int \frac{(3x+2)dx}{(x^2-3x+3)^2} = -\frac{3}{2(x^2-3x+3)} + \frac{13(2x-3)}{6(x^2-3x+3)}$$

$$+ \frac{26}{3\sqrt{3}}\tan^{-1}\frac{2x-3}{\sqrt{3}}$$

$$= \frac{13x-24}{3(x^2-3x+3)} + \frac{26}{3\sqrt{3}}\tan^{-1}\frac{2x-3}{\sqrt{3}}.$$

EXAMPLES.

1. $\int \frac{x^3+x-1}{(x^2+2)^2}dx = -\frac{x-2}{4(x^2+2)} + \frac{1}{2}\log(x^2+2) - \frac{1}{4\sqrt{2}}\tan^{-1}\frac{x}{\sqrt{2}}.$

2. $\int \frac{x^5-x^4+21}{(x^2+3)^3}dx = \frac{4x-9}{4(x^2+3)^2} + \frac{3x+6}{2(x^2+3)} + \frac{1}{2}\log(x^2+3)$

$$+ \frac{1}{2\sqrt{3}}\tan^{-1}\frac{x}{\sqrt{3}}.$$

3. $\int \frac{x^5-2x+1}{x^2(x^2+1)^2}dx = -\frac{3x^2+x+2}{2x(x^2+1)} + \log\frac{(x^2+1)^{\frac{3}{2}}}{x^2} - \frac{3}{2}\tan^{-1}x.$

4. $\int \frac{(4x^2-8x)dx}{(x-1)^2(x^2+1)^2} = \frac{3x^2-x}{(x-1)(x^2+1)} + \log\frac{(x-1)^2}{x^2+1} + \tan^{-1}x.$

5. $\int \dfrac{x^3+8x+21}{(x^2-4x+9)^2}\,dx = \dfrac{3(x-7)}{2(x^2-4x+9)} + \dfrac{1}{2}\log(x^2-4x+9)$

$\qquad + \dfrac{3\sqrt{5}}{2}\tan^{-1}\dfrac{x-2}{\sqrt{5}}.$

6. $\int \dfrac{4x^5(x-1)\,dx}{(x^4+x^2+1)^2} = -\dfrac{2(x^2-1)(x-1)}{3(x^4+x^2+1)} + \log\dfrac{x^2-x+1}{x^2+x+1}$

$\qquad + \dfrac{4}{\sqrt{3}}\left(\tan^{-1}\dfrac{2x+1}{\sqrt{3}} - \dfrac{1}{3}\tan^{-1}\dfrac{2x-1}{\sqrt{3}}\right).$

CHAPTER III.

INTEGRATION BY RATIONALIZATION. INTEGRATION BY SUBSTITUTION.

14. As the preceding chapter provides for the integration of rational fractions, it follows that any rational algebraic function is integrable.

Some irrational expressions may be integrated by substituting a new variable, so related to the old, that the new expression shall be rational.

15. *Expressions involving only fractional powers of x.* Such forms may be rationalized by assuming $x = z^n$, where n is the least common multiple of the denominators of the several fractional exponents.

Take for example, $\int \dfrac{dx}{x^{\frac{1}{2}} + x^{\frac{1}{3}}}.$

Assume $\qquad x = z^6, \ dx = 6z^5 dz;$

then $\qquad x^{\frac{1}{2}} = z^3, \ x^{\frac{1}{3}} = z^2.$

$$\int \frac{dx}{x^{\frac{1}{2}} + x^{\frac{1}{3}}} = \int \frac{6z^5 dz}{z^3 + z^2} = 6\int \frac{z^3 dz}{z+1}.$$

$$\int \frac{z^3 dz}{z+1} = \int \left(z^2 - z + 1 - \frac{1}{z+1}\right) dz$$

$$= \frac{z^3}{3} - \frac{z^2}{2} + z - \log(z+1).$$

Substituting in this, $z = x^{\frac{1}{6}}$, we have

$$\int \frac{dx}{x^{\frac{1}{2}} + x^{\frac{1}{3}}} = 2x^{\frac{1}{2}} - 3x^{\frac{1}{3}} + 6x^{\frac{1}{6}} - 6\log(x^{\frac{1}{6}} + 1).$$

16. *Expressions involving only fractional powers of* $(a+bx)$, *may be rationalized by the method of the preceding article.*

Take for example, $\displaystyle\int \frac{dx}{(x-2)^{\frac{5}{6}}+(x-2)^{\frac{2}{3}}}$.

Assume $x-2=z^6$, $dx=6z^5dz$.

$$\int \frac{dx}{(x-2)^{\frac{5}{6}}+(x-2)^{\frac{2}{3}}} = \int \frac{6z^5 dz}{z^5+z^4} = 6\int \frac{z\,dz}{z+1}$$
$$= 6[z - \log(z+1)].$$

Substituting $z=(x-2)^{\frac{1}{6}}$, we have

$$\int \frac{dx}{(x-2)^{\frac{5}{6}}+(x-2)^{\frac{2}{3}}} = 6(x-2)^{\frac{1}{6}} - 6\log[(x-2)^{\frac{1}{6}}+1].$$

EXAMPLES.

1. $\displaystyle\int \frac{x^{\frac{1}{2}}dx}{x^{\frac{3}{4}}+1} = \frac{4}{3}x^{\frac{3}{4}} - \frac{4}{3}\log(x^{\frac{3}{4}}+1)$.

2. $\displaystyle\int \frac{dx}{x^{\frac{7}{6}}+x^{\frac{4}{3}}} = -\frac{6}{x^{\frac{1}{6}}} + \log \frac{(x^{\frac{1}{6}}+1)^6}{x}$.

3. $\displaystyle\int \frac{x^{\frac{1}{3}}+1}{x^{\frac{7}{6}}+x^{\frac{5}{4}}}dx = -\frac{6}{x^{\frac{1}{6}}} + \frac{12}{x^{\frac{1}{12}}} + 2\log x - 24\log(x^{\frac{1}{12}}+1)$.

4. $\displaystyle\int \frac{dx}{x^{\frac{2}{3}}-x^{\frac{5}{6}}} = \frac{8}{3}x^{\frac{3}{8}} + 2\log \frac{x^{\frac{1}{8}}-1}{x^{\frac{1}{8}}+1} + 4\tan^{-1}x^{\frac{1}{8}}$.

5. $\displaystyle\int \frac{dx}{x\sqrt{x+1}} = \log \frac{\sqrt{x+1}-1}{\sqrt{x+1}+1}$.

6. $\displaystyle\int \frac{x^2 dx}{(4x+1)^{\frac{5}{2}}} = \frac{6x^2+6x+1}{12(4x+1)^{\frac{3}{2}}}$.

7. $\displaystyle\int \frac{dx}{1+\sqrt[3]{x+1}} = \frac{3}{2}(x+1)^{\frac{2}{3}} - 3(x+1)^{\frac{1}{3}} + 3\log(1+\sqrt[3]{x+1})$.

8. $\int \dfrac{dx}{(x^{\frac{1}{2}}+a)^{\frac{1}{2}}} = \dfrac{4}{3}(x^{\frac{1}{2}}-2a)(x^{\frac{1}{2}}+a)^{\frac{1}{2}}.$

9. $\int \dfrac{(2x-1)^{\frac{1}{4}}dx}{x-(2x-1)^{\frac{1}{2}}}$

$= 4(2x-1)^{\frac{1}{4}} - \dfrac{2(2x-1)^{\frac{1}{2}}}{(2x-1)^{\frac{1}{2}}-1} + 3\log\dfrac{(2x-1)^{\frac{1}{4}}-1}{(2x-1)^{\frac{1}{4}}+1}.$

10. $\int \dfrac{x\,dx}{(2x+2)^{\frac{3}{4}}+4(2x+2)^{\frac{1}{4}}}$

$= \dfrac{2}{5}(x+36)(2x+2)^{\frac{1}{4}} - \dfrac{4}{3}(2x+2)^{\frac{3}{4}} - 28\tan^{-1}\left(\dfrac{x+1}{8}\right)^{\frac{1}{4}}.$

17. *Expressions of the form $f(x^2)\cdot x\,dx$, involving fractional powers of $(a+bx^2)$, may also be rationalized by the method of Art. 15.*

Take for example, $\int \dfrac{x^3 dx}{\sqrt{1-x^2}}.$

Assume $1-x^2 = z^2$, $x^2 = 1-z^2$, $x\,dx = -z\,dz.$

$\int \dfrac{x^2 \cdot x\,dx}{\sqrt{1-x^2}} = -\int \dfrac{(1-z^2)z\,dz}{z} = -\int (1-z^2)\,dz$

$= -\left(z - \dfrac{z^3}{3}\right) = -\dfrac{z}{3}(3-z^2) = -\dfrac{\sqrt{1-x^2}}{3}(x^2+2).$

EXAMPLES.

1. $\int \dfrac{x^5 dx}{\sqrt{2x^2+1}} = \dfrac{3x^4 - 2x^2 + 2}{30}\sqrt{2x^2+1}.$

2. $\int x^3(a^2-x^2)^{\frac{1}{3}}dx = \dfrac{5}{132}(6x^4 - a^2x^2 - 5a^4)(a^2-x^2)^{\frac{4}{3}}.$

3. $\int \dfrac{dx}{x\sqrt{x^2+a^2}} = \dfrac{1}{2a}\log\dfrac{\sqrt{x^2+a^2}-a}{\sqrt{x^2+a^2}+a} = \dfrac{1}{a}\log\dfrac{x}{\sqrt{x^2+a^2}+a}.$

4. $\displaystyle\int\frac{xdx}{\sqrt[3]{x^2+1}-1}=\frac{3}{2}\left[\frac{(x^2+1)^{\frac{4}{3}}}{2}+(x^2+1)^{\frac{2}{3}}+\log(\sqrt[3]{x^2+1}-1)\right].$

5. $\displaystyle\int\frac{x\,dx}{x^2+2\sqrt{3-x^2}}=\frac{1}{4}\log(\sqrt{3-x^2}+1)+\frac{3}{4}\log(\sqrt{3-x^2}-3).$

18. *Expressions containing* $\sqrt{x^2+ax+b}$.

If we assume, as in the preceding articles,

$$\sqrt{x^2+ax+b}=z,\quad x^2+ax+b=z^2,$$

the expression for x, and consequently that of dx, in terms of z, will involve radicals. To meet this objection we assume

$$\sqrt{x^2+ax+b}=z-x,\quad ax+b=z^2-2zx,$$

$$x=\frac{z^2-b}{2z+a},\qquad dx=\frac{2(z^2+az+b)dz}{(2z+a)^2},$$

$$\sqrt{x^2+ax+b}=z-x=\frac{z^2+az+b}{2z+a}.$$

Thus $\sqrt{x^2+ax+b}$, x, and dx are expressed rationally in terms of z.

Take for example, $\displaystyle\int\frac{dx}{x\sqrt{x^2-x+2}}.$

Assume $\sqrt{x^2-x+2}=z-x,\quad -x+2=z^2-2zx,$

$$x=\frac{z^2-2}{2z-1},\qquad dx=\frac{2(z^2-z+2)dz}{(2z-1)^2},$$

$$\sqrt{x^2-x+2}=z-x=\frac{z^2-z+2}{2z-1}.$$

$$\therefore\int\frac{dx}{x\sqrt{x^2-x+2}}=\int\frac{2\,dz}{z^2-2}=\frac{1}{\sqrt{2}}\log\frac{z-\sqrt{2}}{z+\sqrt{2}}.$$

Substituting $z=\sqrt{x^2-x+2}+x,$

$$\int\frac{dx}{x\sqrt{x^2-x+2}}=\frac{1}{\sqrt{2}}\log\frac{\sqrt{x^2-x+2}+x-\sqrt{2}}{\sqrt{x^2-x+2}+x+\sqrt{2}}.$$

19. *Expressions containing* $\sqrt{-x^2 + ax + b}$.

To rationalize in this case, it is necessary to resolve $b + ax - x^2$ into two factors. These factors will be real, unless the given radical $\sqrt{b + ax - x^2}$ is imaginary for all values of x. For

$$b + ax - x^2 = \frac{a^2}{4} + b - \left(\frac{a}{2} - x\right)^2$$

$$= \left[\frac{1}{2}(\sqrt{a^2 + 4b} + a) - x\right]\left[\frac{1}{2}(\sqrt{a^2 + 4b} - a) + x\right].$$

These factors are real unless $a^2 + 4b$ is negative, but then $b + ax - x^2$ is negative for all values of x, and consequently $\sqrt{b + ax - x^2}$ is imaginary.

Represent the two factors thus, —

$$b + ax - x^2 = (\alpha - x)(\beta + x).$$

Now assume

$$\sqrt{b + ax - x^2} = \sqrt{(\alpha - x)(\beta + x)} = (\alpha - x)z;$$

then $\quad \beta + x = (\alpha - x)z^2, \quad x = \dfrac{\alpha z^2 - \beta}{z^2 + 1}.$

Thus x is expressed rationally in terms of z.

Take for example, $\displaystyle\int \frac{dx}{x\sqrt{2 + x - x^2}}.$

Assume $\sqrt{2 + x - x^2} = \sqrt{(2 - x)(1 + x)} = (2 - x)z.$

$$1 + x = (2 - x)z^2, \quad x = \frac{2z^2 - 1}{z^2 + 1}, \quad dx = \frac{6z\,dz}{(z^2 + 1)^2}$$

$$\sqrt{2 + x - x^2} = (2 - x)z = \frac{3z}{z^2 + 1}.$$

Therefore,

$$\int \frac{dx}{x\sqrt{2 + x - x^2}} = \int \frac{2\,dz}{2z^2 - 1} = \frac{1}{\sqrt{2}} \log \frac{z\sqrt{2} - 1}{z\sqrt{2} + 1}.$$

Substituting $z = \sqrt{\dfrac{1+x}{2-x}}$,

$$\int \frac{dx}{x\sqrt{2+x-x^2}} = \frac{1}{\sqrt{2}} \log \frac{\sqrt{2+2x}-\sqrt{2-x}}{\sqrt{2+2x}+\sqrt{2-x}}.$$

EXAMPLES.

1. $\displaystyle\int \frac{dx}{x\sqrt{x^2+2x-1}} = 2\tan^{-1}(x+\sqrt{x^2+2x-1})$.

2. $\displaystyle\int \frac{x\,dx}{(2+3x-2x^2)^{\frac{3}{2}}} = \frac{8+6x}{25\sqrt{2+3x-2x^2}}$.

3. $\displaystyle\int \frac{dx}{x^2\sqrt{x^2-2}} = \frac{-1}{x(x+\sqrt{x^2-2})}$ or $= \frac{\sqrt{x^2-2}}{2x}$.

4. $\displaystyle\int \frac{\sqrt{x^2+2x}}{x^2} dx = -\frac{4}{x+\sqrt{x^2+2x}} + \log(x+1+\sqrt{x^2+2x})$

   or $= -2\sqrt{\dfrac{x+2}{x}} + 2\log(\sqrt{x+2}+\sqrt{x})$.

5. $\displaystyle\int \frac{\sqrt{6x-x^2}}{x^2} dx = -2\sqrt{\dfrac{6-x}{x}} + 2\tan^{-1}\sqrt{\dfrac{6-x}{x}}$

   $= -2\sqrt{\dfrac{6-x}{x}} + \cos^{-1}\dfrac{x-3}{3}$.

6. $\displaystyle\int \frac{dx}{(x-1)^2\sqrt{x^2-2x+2}}$

   $= \dfrac{1}{\sqrt{x^2-2x+2}+x} - \dfrac{1}{\sqrt{x^2-2x+2}+x-2}$

   or $= -\dfrac{\sqrt{x^2-2x+2}}{x-1}$.

20. *Integration by Substitution.* This method is used for rationalization, as shown in the preceding articles, but in other cases the introduction of a new variable often simplifies the given expression, and renders it directly integrable. This is illustrated by the following examples.

EXAMPLES.

1. $\int \dfrac{x^2 - x}{(x-2)^3} dx = \log(x-2) - \dfrac{3x-5}{(x-2)^2}.$ Assume $x - 2 = z$.

2. $\int \dfrac{x^3 dx}{(x+1)^4} = \dfrac{18x^2 + 27x + 11}{6(x+1)^3} + \log(x+1).$

    Assume $x + 1 = z$.

3. $\int \dfrac{dx}{x\sqrt{a^2 + x^2}} = \dfrac{1}{a} \log \dfrac{x}{a + \sqrt{a^2 + x^2}}.$ Assume $x = \dfrac{a}{z}$.

4. $\int \dfrac{dx}{x\sqrt{a^2 - x^2}} = \dfrac{1}{a} \log \dfrac{x}{a + \sqrt{a^2 - x^2}}.$ Assume $x = \dfrac{a}{z}$.

5. $\int \dfrac{x^3 dx}{(x^2+1)^{\frac{2}{3}}} = \dfrac{3}{8}(x^2 - 3)(x^2 + 1)^{\frac{1}{3}}.$ Assume $x^2 + 1 = z$.

6. $\int \dfrac{\sin x \, dx}{\sin(x+a)} = (x+a)\cos a - \sin a \log \sin(x+a).$

    Assume $x + a = z$.

7. $\int \dfrac{e^{2x} dx}{(e^x + 1)^{\frac{1}{4}}} = \dfrac{4}{21}(3e^x - 4)(e^x + 1)^{\frac{3}{4}}.$ Assume $e^x + 1 = z$.

8. $\int \dfrac{dx}{e^{2x} - 2e^x} = \dfrac{1}{2e^x} - \dfrac{x}{4} + \dfrac{1}{4}\log(e^x - 2).$ Assume $e^x = z$.

9. $\int \dfrac{(x^2 - 1) dx}{x\sqrt{x^4 + 3x^2 + 1}} = \log \dfrac{x^2 + 1 + \sqrt{x^4 + 3x^2 + 1}}{x}$

    Assume $x + \dfrac{1}{x} = z$.

10. $\int \dfrac{(1 + 2x^n)^{\frac{3}{4}} dx}{x}$

    $= \dfrac{1}{n} \left[\dfrac{4}{3}(1+2x^n)^{\frac{3}{4}} + 2\tan^{-1}(1+2x^n)^{\frac{1}{4}} + \log \dfrac{(1+2x^n)^{\frac{1}{4}} - 1}{(1+2x^n)^{\frac{1}{4}} + 1} \right].$

    Assume $1 + 2x^n = z^4$.

CHAPTER IV.

INTEGRATION BY PARTS. INTEGRATION BY SUCCESSIVE REDUCTION.

21. *Integration by Parts.* From the equation

$$d(uv) = u\,dv + v\,du,$$

we obtain, by integrating both members,

$$uv = \int u\,dv + \int v\,du.$$

Hence $\quad \int u\,dv = uv - \int v\,du.$ . . . . (1)

The use of (1) is called *integration by parts*.
Let us apply it, for example, to

$$\int x \log x\, dx.$$

Let $\quad u = \log x, \quad$ then $\quad dv = x\,dx;$
whence $\quad du = \dfrac{dx}{x}, \quad$ and $\quad v = \dfrac{x^2}{2}.$

Substituting in (1), we have

$$\int \log x \cdot x\,dx = \log x \cdot \frac{x^2}{2} - \int \frac{x^2}{2} \cdot \frac{dx}{x} \quad \ldots \ldots (2)$$
$$= \frac{x^2}{2} \log x - \frac{x^2}{4}.$$

The student should carefully notice how the factors u, dv, v, du, occur in the process, so as to be able to apply it without such a formal substitution as in the preceding example.

On referring to the equation (2), we see that, after selecting for u a certain factor of the given integral, as $\log x$, we obtain the first term in the second member, by integrating as if this factor were constant; also that the expression following the second $\int$, is the same as the preceding term, with the factor $\log x$ replaced by its differential.

Take for another example
$$\int x \cos x\, dx.$$
Assuming $u = \cos x$, we find
$$\int x \cos x\, dx = \cos x \cdot \frac{x^2}{2} - \int \frac{x^2}{2}(-\sin x\, dx).$$

But as the new integral is no simpler than the given one, we gain nothing by this application of the process.

If, however, we let $u = x$, we find
$$\int x \cos x\, dx = x \sin x - \int \sin x\, dx$$
$$= x \sin x + \cos x.$$

о 7.

EXAMPLES.

1. $\int x^2 \log x\, dx = \dfrac{x^3}{3}\left(\log x - \dfrac{1}{3}\right).$

2. $\int x^{n-1} \log x\, dx = \dfrac{x^n}{n}\left(\log x - \dfrac{1}{n}\right).$

3. $\int x \sin x\, dx = -x \cos x + \sin x.$

4. $\int x \log(x+2)\, dx = (x^2 - 4)\log\sqrt{x+2} - \dfrac{x^2}{4} + x.$

5. $\int x e^{ax}\, dx = \dfrac{e^{ax}}{a}\left(x - \dfrac{1}{a}\right).$

6. $\int x \tan^{-1} x\, dx = \dfrac{x^2+1}{2}\tan^{-1} x - \dfrac{x}{2}.$

7. $\int \sin^{-1} x\, dx = x \sin^{-1} x + \sqrt{1 - x^2}.$

8. $\int x \tan^2 x\, dx = x \tan x - \dfrac{x^2}{2} + \log \cos x.$

9. $\int \dfrac{\log(x+1)\,dx}{\sqrt{x+1}} = 2\sqrt{x+1}\,[\log(x+1)-2]$.

10. $\int \dfrac{\log(1+\sqrt{x})\,dx}{\sqrt{x}} = 2(1+\sqrt{x})\log(1+\sqrt{x}) - 2\sqrt{x}$.

11. $\int \tan^{-1}\sqrt{x}\,dx = (1+x)\tan^{-1}\sqrt{x} - \sqrt{x}$.

12. $\int \dfrac{\log x\,dx}{(x+1)^2} = \dfrac{x}{x+1}\log x - \log(x+1)$.

13. $\int x^2 \sin^{-1} x\,dx = \dfrac{x^3}{3}\sin^{-1} x + \dfrac{x^2+2}{9}\sqrt{1-x^2}$.

In each of the following examples integration by parts must be applied successively.

14. $\int x^2 e^x\,dx = (x^2 - 2x + 2)e^x$.

15. $\int x^3 e^{ax}\,dx = \left(x^3 - \dfrac{3x^2}{a} + \dfrac{6x}{a^2} - \dfrac{6}{a^3}\right)\dfrac{e^{ax}}{a}$.

16. $\int x^3 (\log x)^2\,dx = \dfrac{x^4}{4}\left[(\log x)^2 - \dfrac{1}{2}\log x + \dfrac{1}{8}\right]$.

17. $\int \dfrac{(\log x)^2\,dx}{x^{\frac{5}{2}}} = -\dfrac{2}{3x^{\frac{3}{2}}}\left[(\log x)^2 + \dfrac{4}{3}\log x + \dfrac{8}{9}\right]$.

22. Formulæ of Reduction. These are formulæ by which the integral,
$$\int x^m (a+bx^n)^p\,dx,$$
may be made to depend upon a similar integral with either m or p numerically diminished. There are four such formulæ, as follows, —

$\int x^m (a+bx^n)^p\,dx$
$= \dfrac{x^{m-n+1}(a+bx^n)^{p+1}}{(np+m+1)b} - \dfrac{(m-n+1)a}{(np+m+1)b}\int x^{m-n}(a+bx^n)^p\,dx$, (A)

$$\int x^m(a+bx^n)^p dx$$
$$= \frac{x^{m+1}(a+bx^n)^p}{np+m+1} + \frac{anp}{np+m+1}\int x^m(a+bx^n)^{p-1}dx, \ldots \quad (B)$$

$$\int x^m(a+bx^n)^p dx$$
$$= \frac{x^{m+1}(a+bx^n)^{p+1}}{(m+1)a} - \frac{(np+n+m+1)b}{(m+1)a}\int x^{m+n}(a+bx^n)^p dx, \quad (C)$$

$$\int x^m(a+bx^n)^p dx$$
$$= -\frac{x^{m+1}(a+bx^n)^{p+1}}{n(p+1)a} + \frac{np+n+m+1}{n(p+1)a}\int x^m(a+bx^n)^{p+1}dx. \quad (D)$$

Formula (A) changes m into $m-n$.
Formula (B) changes p into $p-1$.
Formula (C) changes m into $m+n$.
Formula (D) changes p into $p+1$.
Formulæ (C) and (D) are used when m or p is negative, requiring an algebraic increase.

23. *Derivation of Formulæ (A) and (C).* Let us put for convenience
$$z = a + bx^n, \quad dz = nbx^{n-1}dx.$$

Then $\int x^m(a+bx^n)^p dx = \int x^m z^p dx = \int x^{m-n+1} z^p x^{n-1} dx.$

Integrating by parts, with $u = x^{m-n+1}$,

$$\int x^m z^p dx = x^{m-n+1}\frac{z^{p+1}}{nb(p+1)} - \frac{m-n+1}{nb(p+1)}\int x^{m-n}z^{p+1}dx.$$

$$nb(p+1)\int x^m z^p dx = x^{m-n+1}z^{p+1} - (m-n+1)\int x^{m-n}z^{p+1}dx. \quad (1)$$

$$\int x^{m-n}z^{p+1}dx = \int (a+bx^n)x^{m-n}z^p dx = a\int x^{m-n}z^p dx + b\int x^m z^p dx. \quad (2)$$

Substituting (2) in (1), and transposing, we have

$$(np+m+1)b \int x^m z^p dx$$
$$= x^{m-n+1}z^{p+1} - (m-n+1)a \int x^{m-n}z^p dx. \quad \ldots \quad (3)$$

Dividing by $(np+m+1)b$, we have (A).
If in (3) we substitute

$$m - n = m', \quad m = m' + n,$$

and transpose, we have

$$(m'+1)a \int x^{m'} z^p dx$$
$$= x^{m'+1}z^{p+1} - (np+m'+n+1)b \int x^{m'+n}z^p dx.$$

Omitting the accents, and dividing by $(m+1)a$, we have (C).

24. *Derivation of Formulæ* (B) *and* (D). If we integrate by parts $\int x^m z^p dx$, calling $u = z^p$, we have

$$\int x^m z^p dx = z^p \frac{x^{m+1}}{m+1} - \frac{nbp}{m+1} \int x^{m+1} z^{p-1} x^{n-1} dx.$$

$$(m+1) \int x^m z^p dx = x^{m+1} z^p - nbp \int x^{m+n} z^{p-1} dx. \quad \ldots \quad (1)$$

$$\int x^m z^p dx = \int (a + bx^n) x^m z^{p-1} dx$$
$$= a \int x^m z^{p-1} dx + b \int x^{m+n} z^{p-1} dx. \quad \ldots \quad (2)$$

Eliminating from (1) and (2), $\int x^{m+n} z^{p-1} dx$, we have

$$(np+m+1) \int x^m z^p dx = x^{m+1} z^p + npa \int x^m z^{p-1} dx. \quad \ldots \quad (3)$$

Dividing by $np+m+1$, we have (B).

FORMULÆ OF REDUCTION. 211

If in (3) we substitute
$$p-1=p', \quad p=p'+1,$$
and transpose, we have
$$n(p'+1)a\int x^m z^{p'} dx = -x^{m+1}z^{p'+1} + (np'+n+m+1)\int x^m z^{p'+1} dx.$$
Omitting the accents, and dividing by $n(p+1)a$, we have (D).

Formulæ (A) and (B) fail, when $np+m+1=0$.
Formula (C) fails, when $m+1=0$.
Formula (D) fails, when $p+1=0$.

EXAMPLES.

1. $\int \dfrac{x^2 dx}{\sqrt{a^2-x^2}} = -\dfrac{x}{2}\sqrt{a^2-x^2} + \dfrac{a^2}{2}\sin^{-1}\dfrac{x}{a}.$

   Here $\int \dfrac{x^2 dx}{\sqrt{a^2-x^2}} = \int x^2(a^2-x^2)^{-\frac{1}{2}} dx.$

   Apply (A), making
   $$m=2, \quad n=2, \quad p=-\dfrac{1}{2}, \quad a=a^2, \quad b=-1.$$
   $$\int x^2(a^2-x^2)^{-\frac{1}{2}} dx = \dfrac{x(a^2-x^2)^{\frac{1}{2}}}{-2} - \dfrac{a^2}{-2}\int (a^2-x^2)^{-\frac{1}{2}} dx$$
   $$= -\dfrac{x}{2}(a^2-x^2)^{\frac{1}{2}} + \dfrac{a^2}{2}\sin^{-1}\dfrac{x}{a}.$$

2. $\int \sqrt{a^2+x^2}\, dx = \dfrac{x}{2}\sqrt{a^2+x^2} + \dfrac{a^2}{2}\log(x+\sqrt{a^2+x^2}).$

   Apply (B), making
   $$m=0, \quad n=2, \quad p=\dfrac{1}{2}, \quad a=a^2, \quad b=1.$$
   $$\int (a^2+x^2)^{\frac{1}{2}} dx = \dfrac{x}{2}(a^2+x^2)^{\frac{1}{2}} + \dfrac{a^2}{2}\int \dfrac{dx}{(a^2+x^2)^{\frac{1}{2}}}$$
   $$= \dfrac{x}{2}(a^2+x^2)^{\frac{1}{2}} + \dfrac{a^2}{2}\log(x+\sqrt{a^2+x^2}).$$

3. $\int \dfrac{dx}{x^3\sqrt{a^2-x^2}} = -\dfrac{\sqrt{a^2-x^2}}{2a^2x^2} + \dfrac{1}{2a^3}\log\dfrac{x}{a+\sqrt{a^2-x^2}}.$

Apply (*C*), making
$$m = -3, \quad n = 2, \quad p = -\frac{1}{2}, \quad a = a^2, \quad b = -1.$$

$$\int x^{-3}(a^2 - x^2)^{-\frac{1}{2}} dx = \frac{x^{-2}(a^2 - x^2)^{\frac{1}{2}}}{-2a^2} - \frac{1}{-2a^2} \int x^{-1}(a^2 - x^2)^{-\frac{1}{2}} dx.$$

Ex. 4, p. 205, gives

$$\int x^{-1}(a^2 - x^2)^{-\frac{1}{2}} dx = \int \frac{dx}{x\sqrt{a^2 - x^2}} = \frac{1}{a} \log \frac{x}{a + \sqrt{a^2 - x^2}}.$$

Substituting, we obtain the complete integral.

4. $\int \dfrac{dx}{(a^2 - x^2)^{\frac{5}{2}}} = \dfrac{(3a^2 - 2x^2)x}{3a^4(a^2 - x^2)^{\frac{3}{2}}}.$

Apply (*D*), making
$$m = 0, \quad n = 2, \quad p = -\frac{5}{2}, \quad a = a^2, \quad b = -1.$$

$$\int (a^2 - x^2)^{-\frac{5}{2}} dx = \frac{x(a^2 - x^2)^{-\frac{3}{2}}}{3a^2} + \frac{2}{3a^2} \int (a^2 - x^2)^{-\frac{3}{2}} dx.$$

Ex. 33, p. 180, gives

$$\int (a^2 - x^2)^{-\frac{3}{2}} dx = \int \frac{dx}{(a^2 - x^2)^{\frac{3}{2}}} = \frac{x}{a^2 \sqrt{a^2 - x^2}}.$$

Substituting this, we have

$$\int \frac{dx}{(a^2 - x^2)^{\frac{5}{2}}} = \frac{x}{3a^2(a^2 - x^2)^{\frac{3}{2}}} + \frac{2x}{3a^4(a^2 - x^2)^{\frac{1}{2}}}$$

$$= \frac{x}{3a^2(a^2 - x^2)^{\frac{3}{2}}} \left[1 + \frac{2(a^2 - x^2)}{a^2} \right]$$

$$= \frac{x}{3a^2(a^2 - x^2)^{\frac{3}{2}}} \cdot \frac{3a^2 - 2x^2}{a^2}.$$

5. $\int \dfrac{x^2 dx}{\sqrt{x^2 + a^2}} = \dfrac{x}{2}\sqrt{x^2 + a^2} - \dfrac{a^2}{2} \log(x + \sqrt{x^2 + a^2}).$

6. $\int (x^2+a^2)^{\frac{3}{2}} dx = \frac{x}{8}(2x^2+5a^2)\sqrt{x^2+a^2} + \frac{3a^4}{8}\log(x+\sqrt{x^2+a^2})$.

7. $\int \sqrt{a^2-x^2}\, dx = \frac{x}{2}\sqrt{a^2-x^2} + \frac{a^2}{2}\sin^{-1}\frac{x}{a}$.

8. $\int \sqrt{\frac{a-x}{b+x}}\, dx = \sqrt{(a-x)(b+x)} + (a+b)\sin^{-1}\sqrt{\frac{x+b}{a+b}}$.

Substitute $b+x=z^2$, and the integral takes the form of Ex. 7.

9. $\int \sqrt{\frac{x+a}{x+b}}\, dx$
$= \sqrt{(x+a)(x+b)} + (a-b)\log(\sqrt{x+a}+\sqrt{x+b})$.

Substitute $x+b=z^2$, and the integral takes the form of Ex. 2.

10. $\int \sqrt{2ax-x^2}\, dx = \frac{x-a}{2}\sqrt{2ax-x^2} + \frac{a^2}{2}\text{vers}^{-1}\frac{x}{a}$.

$\int \sqrt{2ax-x^2}\, dx = \int \sqrt{a^2-(x-a)^2}\, dx$,

which is in the form of Ex. 7.

11. $\int x^2\sqrt{a^2-x^2}\, dx = \frac{x}{8}(2x^2-a^2)\sqrt{a^2-x^2} + \frac{a^4}{8}\sin^{-1}\frac{x}{a}$.

12. $\int x^2\sqrt{x^2+a^2}\, dx = \frac{x}{8}(2x^2+a^2)\sqrt{x^2+a^2} - \frac{a^4}{8}\log(x+\sqrt{x^2+a^2})$.

13. $\int (a^2-x^2)^{\frac{3}{2}} dx = \frac{x}{8}(5a^2-2x^2)\sqrt{a^2-x^2} + \frac{3a^4}{8}\sin^{-1}\frac{x}{a}$.

14. $\int \frac{dx}{(a^2-x^2)^{\frac{5}{2}}} = \frac{(15a^4-20a^2x^2+8x^4)x}{15a^6(a^2-x^2)^{\frac{3}{2}}}$

15. $\int \frac{dx}{(x^2+a^2)^n}$
$= \frac{1}{2(n-1)a^2}\left[\frac{x}{(x^2+a^2)^{n-1}} + (2n-3)\int \frac{dx}{(x^2+a^2)^{n-1}}\right]$.

16. $\int \dfrac{dx}{x^4(x^2+2)} = -\dfrac{1}{6x^3} + \dfrac{1}{4x} + \dfrac{1}{4\sqrt{2}}\tan^{-1}\dfrac{x}{\sqrt{2}}$.

17. $\int \dfrac{x^5 dx}{\sqrt{1-x^3}} = -\dfrac{2}{45}(3x^6+4x^3+8)\sqrt{1-x^3}$.

18. $\int \dfrac{dx}{x^4\sqrt{1-x^2}} = -\dfrac{2x^2+1}{3x^3}\sqrt{1-x^2}$.

19. $\int \dfrac{x\,dx}{\sqrt{2ax-x^2}} = -\sqrt{2ax-x^2} + a\operatorname{vers}^{-1}\dfrac{x}{a}$.

Here $\int \dfrac{x\,dx}{\sqrt{2ax-x^2}} = \int \dfrac{x^{\frac{1}{2}} dx}{\sqrt{2a-x}}$.

Apply (A), and the integral is reduced to

$$\int \dfrac{dx}{\sqrt{2ax-x^2}} = \operatorname{vers}^{-1}\dfrac{x}{a}.$$

20. $\int \dfrac{dx}{x\sqrt{2ax-x^2}} = -\dfrac{\sqrt{2ax-x^2}}{ax}$.

21. $\int \dfrac{x^m dx}{\sqrt{2ax-x^2}} = -\dfrac{x^{m-1}\sqrt{2ax-x^2}}{m} + \dfrac{(2m-1)a}{m}\int \dfrac{x^{m-1}dx}{\sqrt{2ax-x^2}}$.

22. $\int \dfrac{dx}{x^m\sqrt{2ax-x^2}}$

$= -\dfrac{\sqrt{2ax-x^2}}{(2m-1)ax^m} + \dfrac{m-1}{(2m-1)a}\int \dfrac{dx}{x^{m-1}\sqrt{2ax-x^2}}$.

23. $\int x^m\sqrt{2ax-x^2}\,dx$

$= -\dfrac{x^{m-1}(2ax-x^2)^{\frac{3}{2}}}{m+2} + \dfrac{(2m+1)a}{m+2}\int x^{m-1}\sqrt{2ax-x^2}\,dx$.

24. $\int \dfrac{\sqrt{2ax-x^2}\,dx}{x^m}$

$= -\dfrac{(2ax-x^2)^{\frac{3}{2}}}{(2m-3)ax^m} + \dfrac{m-3}{(2m-3)a}\int \dfrac{\sqrt{2ax-x^2}\,dx}{x^{m-1}}$.

CHAPTER V.

TRIGONOMETRIC INTEGRALS.

25. *Required* $\int \tan^n x \, dx$, or $\int \cot^n x \, dx$.

These forms can be readily integrated when n is an integer, positive or negative.

$$\int \tan^n x \, dx = \int \tan^{n-2} x (\sec^2 x - 1) \, dx$$
$$= \int \tan^{n-2} x \sec^2 x \, dx - \int \tan^{n-2} x \, dx$$
$$= \frac{\tan^{n-1} x}{n-1} - \int \tan^{n-2} x \, dx.$$

Thus $\int \tan^n x \, dx$ is made to depend upon $\int \tan^{n-2} x \, dx$, and ultimately, by successive reductions, upon $\int \tan x \, dx$ or $\int dx$.

When n is negative, the integral takes the form

$$\int \cot^n x \, dx,$$

which can be integrated in a similar manner.

For example, required $\int \tan^5 x \, dx$.

$$\int \tan^5 x \, dx = \int \tan^3 x (\sec^2 x - 1) \, dx$$
$$= \frac{\tan^4 x}{4} - \int \tan^3 x \, dx.$$

$$\int \tan^3 x \, dx = \int \tan x (\sec^2 x - 1) \, dx$$
$$= \frac{\tan^2 x}{2} - \log \sec x.$$

Hence $\quad \int \tan^5 x \, dx = \dfrac{\tan^4 x}{4} - \dfrac{\tan^2 x}{2} + \log \sec x.$

26. *Required* $\int \sec^n x\, dx$, or $\int \operatorname{cosec}^n x\, dx$.

These forms can be readily integrated, when n is an even positive integer.

$$\int \sec^n x\, dx = \int \sec^{n-2} x \sec^2 x\, dx$$
$$= \int (\tan^2 x + 1)^{\frac{n-2}{2}} \sec^2 x\, dx.$$

If n is even, $\dfrac{n-2}{2}$ will be a whole number, and the first factor can be expanded by the Binomial Theorem, and the terms integrated directly.

The following example will illustrate the process.

$$\int \sec^6 x\, dx = \int \sec^4 x \sec^2 x\, dx$$
$$= \int (\tan^2 x + 1)^2 \sec^2 x\, dx = \int (\tan^4 x + 2\tan^2 x + 1)\sec^2 x\, dx$$
$$= \frac{\tan^5 x}{5} + \frac{2\tan^3 x}{3} + \tan x.$$

27. *Required* $\int \tan^m x \sec^n x\, dx$, or $\int \cot^m x \operatorname{cosec}^n x\, dx$.

These forms may be readily integrated when n is a positive even number, or when m is a positive odd number.

When n is even, the method of Art. 26 is applicable.

This is illustrated by the following example:

$$\int \tan^6 x \sec^4 x\, dx = \int \tan^6 x (\tan^2 x + 1)\sec^2 x\, dx$$
$$= \int (\tan^8 x + \tan^6 x)\sec^2 x\, dx = \frac{\tan^9 x}{9} + \frac{\tan^7 x}{7}.$$

When m is odd, proceed as follows:

$$\int \tan^m x \sec^n x\, dx = \int \tan^{m-1} x \sec^{n-1} x \sec x \tan x\, dx$$
$$= \int (\sec^2 x - 1)^{\frac{m-1}{2}} \sec^{n-1} x \sec x \tan x\, dx.$$

Since m is odd, $\dfrac{m-1}{2}$ will be a whole number, and the first factor can then be expanded, and the terms integrated separately.

The following example illustrates the process.

$$\int \tan^5 x \sec^3 x\, dx = \int \tan^4 x \sec^2 x \sec x \tan x\, dx$$

$$= \int (\sec^2 x - 1)^2 \sec^2 x \sec x \tan x\, dx$$

$$= \int (\sec^6 x - 2\sec^4 x + \sec^2 x) \sec x \tan x\, dx$$

$$= \frac{\sec^7 x}{7} - \frac{2\sec^5 x}{5} + \frac{\sec^3 x}{3}.$$

EXAMPLES.

1. $\displaystyle\int \tan^3 x\, dx = \frac{\tan^2 x}{2} + \log \cos x.$

2. $\displaystyle\int \tan^6 x\, dx = \frac{\tan^5 x}{5} - \frac{\tan^3 x}{3} + \tan x - x.$

3. $\displaystyle\int \cot^4 \frac{x}{3}\, dx = -\cot^3 \frac{x}{3} + 3\cot \frac{x}{3} + x.$

4. $\displaystyle\int \tan^5 \frac{x}{4}\, dx = \tan^4 \frac{x}{4} - 2\tan^2 \frac{x}{4} + \log \sec^4 \frac{x}{4}.$

5. $\displaystyle\int \cot^7 x\, dx = -\frac{\cot^6 x}{6} + \frac{\cot^4 x}{4} - \frac{\cot^2 x}{2} - \log \sin x.$

6. When n is even,

$$\int \tan^n x\, dx = \frac{\tan^{n-1} x}{n-1} - \frac{\tan^{n-3} x}{n-3} + \frac{\tan^{n-5} x}{n-5} - \cdots$$
$$+ (-1)^{\frac{n+2}{2}} (\tan x - x).$$

7. When n is odd,

$$\int \tan^n x\, dx = \frac{\tan^{n-1}x}{n-1} - \frac{\tan^{n-3}x}{n-3} + \frac{\tan^{n-5}x}{n-5} - \cdots$$
$$+ (-1)^{\frac{n+1}{2}}\left(\frac{\tan^2 x}{2} - \log \sec x\right).$$

8. $\int \sec^8 x\, dx = \frac{\tan^7 x}{7} + \frac{3\tan^5 x}{5} + \tan^3 x + \tan x.$

9. $\int \cosec^6 2x\, dx = -\frac{\cot^5 2x}{10} - \frac{\cot^3 2x}{3} - \frac{\cot 2x}{2}.$

10. $\int \tan^4 x \sec^4 x\, dx = \frac{\tan^7 x}{7} + \frac{\tan^5 x}{5}.$

11. $\int \frac{\sec^6 x\, dx}{\tan^4 x} = \tan x - 2\cot x - \frac{\cot^3 x}{3}.$

12. $\int \tan^{\frac{3}{2}} x \sec^4 x\, dx = \frac{2\tan^{\frac{5}{2}}x}{5} + \frac{2\tan^{\frac{9}{2}}x}{9}.$

13. $\int \cot^5 x \cosec^4 x\, dx = -\frac{\cot^6 x}{6} - \frac{\cot^8 x}{8}.$

14. $\int \tan^3 x \sec^5 x\, dx = \frac{\sec^7 x}{7} - \frac{\sec^5 x}{5}.$

15. $\int \cot^5 x \cosec^5 x\, dx = -\frac{\cosec^9 x}{9} + \frac{2\cosec^7 x}{7} - \frac{\cosec^5 x}{5}.$

16. $\int \tan^5 x \sec^{\frac{3}{2}} x\, dx = 2\sec^{\frac{3}{2}} x\left(\frac{\sec^4 x}{11} - \frac{2\sec^2 x}{7} + \frac{1}{3}\right).$

17. $\int (\tan x + \cot x)^3\, dx = \frac{1}{2}(\tan^2 x - \cot^2 x) + \log \tan^2 x.$

18. $\int \frac{\sec^{10} x + 1}{\sec^2 x + 1}\, dx = \frac{\tan^7 x}{7} + \frac{2\tan^5 x}{5} + \frac{2\tan^3 x}{3} + x.$

19. $\int (\sec x + \tan x)^4\, dx = \frac{8}{3}(\sec^3 x + \tan^3 x) - 4\sec x + x.$

TRIGONOMETRIC INTEGRALS.

28. *Required* $\int \sin^m x \cos^n x \, dx$.

This is readily integrated when m or n is a positive odd number, or when $m + n$ is a negative even number.

Suppose n to be odd and positive.

$$\int \sin^m x \cos^n x \, dx = \int \sin^m x (1 - \sin^2 x)^{\frac{n-1}{2}} \cos x \, dx.$$

As $\dfrac{n-1}{2}$ is a positive integer, the second factor can be expanded, and the terms integrated separately.

For example,

$$\int \sin^2 x \cos^5 x \, dx = \int \sin^2 x (1 - \sin^2 x)^2 \cos x \, dx$$

$$= \int (\sin^6 x - 2\sin^4 x + \sin^2 x) \cos x \, dx$$

$$= \frac{\sin^7 x}{7} - \frac{2\sin^5 x}{5} + \frac{\sin^3 x}{3}.$$

A similar process may be used, when m is odd and positive. For example,

$$\int \sin^3 x \cos^2 x \, dx = \int \cos^2 x (1 - \cos^2 x) \sin x \, dx$$

$$= \int (\cos^2 x - \cos^4 x) \sin x \, dx$$

$$= -\frac{\cos^3 x}{3} + \frac{\cos^5 x}{5}.$$

When $m + n$ is a negative even number, the form can be integrated by expressing it in terms of $\sec x$ and $\tan x$. Thus

$$\int \sin^m x \cos^n x \, dx = \int \frac{\sin^m x}{\cos^m x} \cos^{m+n} x \, dx$$

$$= \int \tan^m x \sec^{-m-n} x \, dx.$$

Since $-m-n$ is positive and even, the method of Art. 27 is applicable.

For example, consider $\int \dfrac{\sin^2 x}{\cos^4 x} dx$.

Here $m=2, \quad n=-4, \quad m+n=-2.$

$$\int \dfrac{\sin^2 x}{\cos^4 x} dx = \int \tan^2 x \sec^2 x \, dx = \dfrac{\tan^3 x}{3}.$$

EXAMPLES.

1. $\int \sin^4 x \cos^3 x \, dx = \dfrac{\sin^5 x}{5} - \dfrac{\sin^7 x}{7}.$

2. $\int \sin^5 x \cos^4 x \, dx = -\dfrac{\cos^5 x}{5} + \dfrac{2\cos^7 x}{7} - \dfrac{\cos^9 x}{9}.$

3. $\int \sin^7 x \, dx = \dfrac{\cos^7 x}{7} - \dfrac{3\cos^5 x}{5} + \cos^3 x - \cos x.$

4. $\int \cos^5 \dfrac{x}{5} dx = \sin^5 \dfrac{x}{5} - \dfrac{10}{3}\sin^3 \dfrac{x}{5} + 5 \sin \dfrac{x}{5}.$

5. $\int \dfrac{\cos^5 x \, dx}{\sin^2 x} = \dfrac{\sin^3 x}{3} - 2 \sin x - \dfrac{1}{\sin x}.$

6. $\int \sin^5 x \sqrt[3]{\cos x} \, dx = -3 \sqrt[3]{\cos x}\left(\dfrac{\cos x}{4} - \dfrac{\cos^3 x}{5} + \dfrac{\cos^5 x}{16}\right).$

7. $\int \dfrac{\cos^4 x \, dx}{\sin^6 x} = -\dfrac{\cot^5 x}{5}.$

8. $\int \dfrac{dx}{\sin^2 x \cos^4 x} = \dfrac{\tan^3 x}{3} + 2 \tan x - \cot x.$

9. $\int \dfrac{dx}{\sin^3 x \cos^5 x} = \dfrac{\tan^4 x}{4} + \dfrac{3\tan^2 x}{2} - \dfrac{\cot^2 x}{2} + 3 \log \tan x.$

10. $\int \dfrac{\sin^{\frac{2}{3}} x \, dx}{\cos^{\frac{8}{3}} x} = \dfrac{3 \tan^{\frac{5}{3}} x}{5}.$

11. $\int \dfrac{dx}{\sqrt{\sin^3 x \cos^5 x}} = \dfrac{2\sqrt{\tan x}}{3}(\tan x - 3 \cot x).$

12. $\int (\sin x + \cos x)^5 dx = \sin x - \cos x + \dfrac{8}{3}(\sin^3 x - \cos^3 x)$

$\qquad\qquad\qquad\qquad\qquad - \dfrac{4}{5}(\sin^5 x - \cos^5 x).$

13. $\int \left(\dfrac{1}{\sin x} - \dfrac{1}{\cos x}\right)^4 dx = \dfrac{1}{3}(\tan^3 x - \cot^3 x) - 2(\tan^2 x - \cot^2 x)$

$\qquad\qquad\qquad\qquad + 7(\tan x - \cot x) - 8 \log \tan x.$

14. $\int \dfrac{\sin^{n-3} x}{\cos^{n+3} x}(\sin x \cos x - 1)^2 dx = \dfrac{\tan^{n+2} x}{n+2} - \dfrac{2\tan^{n+1} x}{n+1} + \dfrac{3\tan^n x}{n}$

$\qquad\qquad\qquad\qquad\qquad - \dfrac{2\tan^{n-1} x}{n-1} + \dfrac{\tan^{n-2} x}{n-2}.$

29. *Integration by Multiple Angles.* By means of the proper formulæ of trigonometry, $\sin^m x \cos^n x$, when m and n are positive integers, can be expressed in a series of terms of the first degree, involving sines and cosines of multiples of x.

If we use the method of Art. 28 for integrating terms with odd exponents, occurring during the process, the following formulæ for the double angle will be sufficient for the transformation of the terms with even exponents.

$$2 \sin x \cos x = \sin 2x,$$
$$2 \sin^2 x = 1 - \cos 2x,$$
$$2 \cos^2 x = 1 + \cos 2x.$$

For example, required

$$\int \sin^4 x \cos^2 x \, dx.$$

$\sin^4 x \cos^2 x = (\sin x \cos x)^2 \sin^2 x = \dfrac{1}{8} \sin^2 2x (1 - \cos 2x)$

$\qquad\qquad = -\dfrac{1}{8} \sin^2 2x \cos 2x + \dfrac{1}{16}(1 - \cos 4x).$

Hence $\int \sin^4 x \cos^2 x \, dx = -\dfrac{\sin^3 2x}{48} + \dfrac{x}{16} - \dfrac{\sin 4x}{64}.$

EXAMPLES.

1. $\displaystyle\int \sin^4 x\, dx = \frac{1}{4}\left(\frac{3x}{2} - \sin 2x + \frac{\sin 4x}{8}\right).$

2. $\displaystyle\int \cos^4 x\, dx = \frac{1}{4}\left(\frac{3x}{2} + \sin 2x + \frac{\sin 4x}{8}\right).$

3. $\displaystyle\int \sin^2 x \cos^2 x\, dx = \frac{1}{8}\left(x - \frac{\sin 4x}{4}\right).$

4. $\displaystyle\int \sin^6 x\, dx = \frac{1}{16}\left(5x - 4\sin 2x + \frac{\sin^3 2x}{3} + \frac{3}{4}\sin 4x\right).$

5. $\displaystyle\int \cos^6 x\, dx = \frac{1}{16}\left(5x + 4\sin 2x - \frac{\sin^3 2x}{3} + \frac{3}{4}\sin 4x\right).$

6. $\displaystyle\int \sin^4 x \cos^4 x\, dx = \frac{1}{128}\left(3x - \sin 4x + \frac{\sin 8x}{8}\right).$

7. $\displaystyle\int \cos^6 x \sin^2 x\, dx = \frac{1}{128}\left(5x + \frac{8}{3}\sin^3 2x - \sin 4x - \frac{\sin 8x}{8}\right).$

8. $\displaystyle\int \sin^8 x\, dx$
$= \frac{1}{16}\left(\frac{35x}{8} - 4\sin 2x + \frac{2}{3}\sin^3 2x + \frac{7}{8}\sin 4x + \frac{\sin 8x}{64}\right).$

30. Integration of Trigonometric Functions by Transformation into Algebraic Functions.

If in the integral $\int \sin^m x \cos^n x\, dx$, we assume $\sin x = z$, we have also
$$\cos x = (1 - z^2)^{\frac{1}{2}}, \quad x = \sin^{-1} z, \quad dx = \frac{dz}{\sqrt{1 - z^2}}.$$

Hence $\displaystyle\int \sin^m x \cos^n x\, dx = \int z^m (1 - z^2)^{\frac{n}{2}} \frac{dz}{\sqrt{1 - z^2}}$
$$= \int z^m (1 - z^2)^{\frac{n-1}{2}} dz.$$

By means of the formulæ of reduction, this form is integrable for all integral values of m and n, positive or negative.

In the preceding transformation we might have assumed $\cos x = z$, instead of $\sin x = z$.

Any expression containing $\sin x$ and $\cos x$, free from radicals, can thus be integrated, either by a formula of reduction or by rationalization. Moreover, since the other trigonometric functions can be expressed rationally in terms of the sine and cosine, it follows that any rational trigonometric expression can be integrated.

EXAMPLES.

1. $\displaystyle\int \sin^2 x \cos^4 x \, dx = \left(\frac{\cos x}{8} + \frac{\cos^3 x}{12} - \frac{\cos^5 x}{3}\right)\frac{\sin x}{2} + \frac{x}{16}.$

   Assume $\cos x = z$, $\sin^2 x = 1 - z^2$, $dx = -\dfrac{dz}{\sqrt{1-z^2}}.$

   $\displaystyle\int \sin^2 x \cos^4 x \, dx = -\int z^4 (1-z^2)^{\frac{1}{2}} dz.$

   By the formulæ of reduction,

   $\displaystyle\int z^4 (1-z^2)^{\frac{1}{2}} dz = \frac{1}{2}\left(\frac{z^5}{3} - \frac{z^3}{12} - \frac{z}{8}\right)(1-z^2)^{\frac{1}{2}} - \frac{1}{16}\cos^{-1} z.$

   Substituting $z = \cos x$, we have the integral required.

2. $\displaystyle\int \sec^3 x \, dx = \frac{\sec x \tan x}{2} + \frac{1}{2}\log(\sec x + \tan x).$

   Assume $\sec x = z$, $x = \sec^{-1} z$, $dx = \dfrac{dz}{z\sqrt{z^2-1}}.$

   $\displaystyle\int \sec^3 x \, dx = \int \frac{z^2 \, dz}{\sqrt{z^2-1}} = \frac{z}{2}\sqrt{z^2-1} + \frac{1}{2}\log(z + \sqrt{z^2-1})$

   $\displaystyle = \frac{\sec x \tan x}{2} + \frac{1}{2}\log(\sec x + \tan x).$

3. $\displaystyle\int \frac{dx}{\sin x \cos^2 x} = \frac{1}{\cos x} + \log \tan \frac{x}{2}.$

4. $\int \dfrac{dx}{\sin^2 x \cos^3 x} = \dfrac{\sin x}{2\cos^2 x} - \dfrac{1}{\sin x} + \dfrac{3}{2}\log(\sec x + \tan x).$

5. $\int \dfrac{\cos^4 x\, dx}{\sin^3 x} = -\dfrac{\cos x}{2\sin^2 x} - \cos x - \dfrac{3}{2}\log \tan \dfrac{x}{2}.$

6. $\int \dfrac{\sin^2 x\, dx}{\cos^5 x} = \dfrac{\sin^3 x}{4\cos^4 x} + \dfrac{\sin x}{8\cos^2 x} - \dfrac{1}{8}\log(\sec x + \tan x).$

Assume $\tan x = z$.

7. $\int \dfrac{dx}{\tan^2 x - 1} = \dfrac{1}{4}\log \tan\left(x - \dfrac{\pi}{4}\right) - \dfrac{x}{2}.$

8. $\int \dfrac{\tan(x+a)\,dx}{\tan x} = x - \tan a \log(\cot x - \tan a).$

9. $\int \dfrac{dx}{a \tan x + b} = \dfrac{bx}{a^2 + b^2} + \dfrac{a}{a^2 + b^2}\log(a \sin x + b \cos x).$

31. *Trigonometric Formulæ of Reduction.*

By means of the following formulæ, $\int \sin^m x \cos^n x\, dx$ may be obtained for all integral values of m and n, by successive reduction.

$\int \sin^m x \cos^n x\, dx$
$$= -\dfrac{\sin^{m-1} x \cos^{n+1} x}{m+n} + \dfrac{m-1}{m+n}\int \sin^{m-2} x \cos^n x\, dx. \quad (1)$$

$$\int \dfrac{\cos^n x\, dx}{\sin^m x} = -\dfrac{\cos^{n+1} x}{(m-1)\sin^{m-1} x} + \dfrac{m-n-2}{m-1}\int \dfrac{\cos^n x\, dx}{\sin^{m-2} x}. \quad (2)$$

$\int \sin^m x \cos^n x\, dx$
$$= \dfrac{\sin^{m+1} x \cos^{n-1} x}{m+n} + \dfrac{n-1}{m+n}\int \sin^m x \cos^{n-2} x\, dx. \quad (3)$$

$$\int \dfrac{\sin^m x\, dx}{\cos^n x} = \dfrac{\sin^{m+1} x}{(n-1)\cos^{n-1} x} + \dfrac{n-m-2}{n-1}\int \dfrac{\sin^m x\, dx}{\cos^{n-2} x}. \quad (4)$$

$$\int \sin^m x\, dx = -\dfrac{\sin^{m-1} x \cos x}{m} + \dfrac{m-1}{m}\int \sin^{m-2} x\, dx. \quad (5)$$

$$\int \frac{dx}{\sin^m x} = -\frac{\cos x}{(m-1)\sin^{m-1} x} + \frac{m-2}{m-1}\int \frac{dx}{\sin^{m-2} x}. \quad \ldots \quad (6)$$

$$\int \cos^n x \, dx = \frac{\sin x \cos^{n-1} x}{n} + \frac{n-1}{n}\int \cos^{n-2} x \, dx. \quad \ldots \quad (7)$$

$$\int \frac{dx}{\cos^n x} = \frac{\sin x}{(n-1)\cos^{n-1} x} + \frac{n-2}{n-1}\int \frac{dx}{\cos^{n-2} x}. \quad \ldots \quad (8)$$

32. Derivation of the Formulæ in the preceding article.

To derive (1), we integrate by parts with $u = \sin^{m-1} x$.

$$\int \sin^m x \cos^n x \, dx = -\frac{\sin^{m-1} x \cos^{n+1} x}{n+1} + \frac{m-1}{n+1}\int \sin^{m-2} x \cos^{n+2} x \, dx.$$

$$\int \sin^{m-2} x \cos^{n+2} x \, dx = \int \sin^{m-2} x \cos^n x \, dx - \int \sin^m x \cos^n x \, dx.$$

Substituting this in the preceding equation, we have

$$(m+n)\int \sin^m x \cos^n x \, dx$$
$$= -\sin^{m-1} x \cos^{n+1} x + (m-1)\int \sin^{m-2} x \cos^n x \, dx,$$

which gives (1).

To derive (2), substitute in (1)

$$m - 2 = -m', \quad m = 2 - m',$$

afterwards omitting the accents and transposing.

To derive (3), integrate by parts with $u = \cos^{n-1} x$, and proceed as in the derivation of (1).

To derive (4), substitute in (3)

$$n - 2 = -n', \quad n = 2 - n',$$

afterwards omitting the accents and transposing.

To derive (5) and (6), make $n = 0$ in (1) and (2), respectively.
To derive (7) and (8), make $m = 0$ in (3) and (4), respectively.

EXAMPLES.

1. $\int \sin^6 x\, dx = -\dfrac{\cos x}{2}\left(\dfrac{\sin^5 x}{3} + \dfrac{5}{12}\sin^3 x + \dfrac{5}{8}\sin x\right) + \dfrac{5x}{16}$.

2. $\int \operatorname{cosec}^5 x\, dx = -\dfrac{\cos x}{4}\left(\dfrac{1}{\sin^4 x} + \dfrac{3}{2\sin^2 x}\right) + \dfrac{3}{8}\log\tan\dfrac{x}{2}$.

3. $\int \sec^7 x\, dx = \dfrac{\sin x}{2\cos^2 x}\left(\dfrac{1}{3\cos^4 x} + \dfrac{5}{12\cos^2 x} + \dfrac{5}{8}\right)$
$\qquad\qquad + \dfrac{5}{16}\log(\sec x + \tan x)$.

4. $\int \cos^8 x\, dx = \dfrac{\sin x}{8}\left(\cos^7 x + \dfrac{7}{6}\cos^5 x + \dfrac{35}{24}\cos^3 x + \dfrac{35}{16}\cos x\right) + \dfrac{35x}{128}$.

5. $\int \sin^4 x \cos^2 x\, dx = \dfrac{\cos x}{2}\left(\dfrac{\sin^5 x}{3} - \dfrac{\sin^3 x}{12} - \dfrac{\sin x}{8}\right) + \dfrac{x}{16}$.

6. $\int \dfrac{\cos^4 x\, dx}{\sin^2 x} = -\dfrac{\cos x}{\sin x}\left(\cos^4 x + \sin^2 x\cos^2 x + \dfrac{3}{2}\sin^2 x\right) - \dfrac{3x}{2}$
$\qquad = -\dfrac{\cos x}{2\sin x}(3 - \cos^2 x) - \dfrac{3x}{2}$.

7. $\int \dfrac{dx}{\sin^4 x \cos^3 x} = -\dfrac{1}{\cos^2 x}\left(\dfrac{1}{3\sin^3 x} + \dfrac{5}{3\sin x} - \dfrac{5}{2}\sin x\right)$
$\qquad\qquad + \dfrac{5}{2}\log(\sec x + \tan x)$.

33. Required $\int \dfrac{dx}{a + b\sin x}$.

$$a + b\sin x = a\left(\cos^2\dfrac{x}{2} + \sin^2\dfrac{x}{2}\right) + 2b\sin\dfrac{x}{2}\cos\dfrac{x}{2}.$$

$$\int \dfrac{dx}{a + b\sin x} = \int \dfrac{\sec^2\dfrac{x}{2}\, dx}{a + 2b\tan\dfrac{x}{2} + a\tan^2\dfrac{x}{2}} = \int \dfrac{a\sec^2\dfrac{x}{2}\, dx}{\left(a\tan\dfrac{x}{2} + b\right)^2 + a^2 - b^2}$$

$$= 2\int \dfrac{dz}{z^2 + a^2 - b^2}, \text{ where } z = a\tan\dfrac{x}{2} + b.$$

TRIGONOMETRIC INTEGRALS.

If $a > b$, numerically,

$$\int \frac{dx}{a + b\sin x} = \frac{2}{\sqrt{a^2 - b^2}} \tan^{-1} \frac{z}{\sqrt{a^2 - b^2}}$$

$$= \frac{2}{\sqrt{a^2 - b^2}} \tan^{-1} \frac{a\tan\frac{x}{2} + b}{\sqrt{a^2 - b^2}}.$$

If $a < b$, numerically,

$$\int \frac{dx}{a + b\sin x} = 2\int \frac{dz}{z^2 - (b^2 - a^2)} = \frac{1}{\sqrt{b^2 - a^2}} \log \frac{z - \sqrt{b^2 - a^2}}{z + \sqrt{b^2 - a^2}}$$

$$= \frac{1}{\sqrt{b^2 - a^2}} \log \frac{a\tan\frac{x}{2} + b - \sqrt{b^2 - a^2}}{a\tan\frac{x}{2} + b + \sqrt{b^2 - a^2}}.$$

34. Required $\int \frac{dx}{a + b\cos x}$.

$$a + b\cos x = a\left(\cos^2\frac{x}{2} + \sin^2\frac{x}{2}\right) + b\left(\cos^2\frac{x}{2} - \sin^2\frac{x}{2}\right)$$

$$= (a + b)\cos^2\frac{x}{2} + (a - b)\sin^2\frac{x}{2}.$$

$$\int \frac{dx}{a + b\cos x} = \int \frac{\sec^2\frac{x}{2}\,dx}{a + b + (a - b)\tan^2\frac{x}{2}}.$$

If we put $\tan\frac{x}{2} = z$,

$$\int \frac{dx}{a + b\cos x} = 2\int \frac{dz}{a + b + (a - b)z^2} = \frac{2}{a - b}\int \frac{dz}{z^2 + \frac{a + b}{a - b}}.$$

If $a > b$, numerically,

$$\int \frac{dx}{a + b\cos x} = \frac{2}{a - b}\sqrt{\frac{a - b}{a + b}} \tan^{-1} \frac{z\sqrt{a - b}}{\sqrt{a + b}}$$

$$= \frac{2}{\sqrt{a^2 - b^2}} \tan^{-1}\left(\sqrt{\frac{a - b}{a + b}} \tan\frac{x}{2}\right).$$

If $a < b$, numerically,

$$\int \frac{dx}{a + b\cos x} = -\frac{2}{b-a}\int \frac{dz}{z^2 - \frac{b+a}{b-a}}$$

$$= -\frac{1}{\sqrt{b^2 - a^2}}\log \frac{z\sqrt{b-a} - \sqrt{b+a}}{z\sqrt{b-a} + \sqrt{b+a}}$$

$$= \frac{1}{\sqrt{b^2 - a^2}}\log \frac{\sqrt{b-a}\tan\frac{x}{2} + \sqrt{b+a}}{\sqrt{b-a}\tan\frac{x}{2} - \sqrt{b+a}}.$$

35. *Required* $\int e^{ax}\sin nx\, dx$, *and* $\int e^{ax}\cos nx\, dx$.

Integrating by parts, with $u = e^{ax}$,

$$\int e^{ax}\sin nx\, dx = -\frac{e^{ax}\cos nx}{n} + \frac{a}{n}\int e^{ax}\cos nx\, dx. \quad . \quad (1)$$

Integrating the same, with $u = \sin nx$,

$$\int e^{ax}\sin nx\, dx = \frac{e^{ax}\sin nx}{a} - \frac{n}{a}\int e^{ax}\cos nx\, dx. \quad . \quad . \quad (2)$$

Eliminating from (1) and (2) $\int e^{ax}\cos nx\, dx$, we have

$$(a^2 + n^2)\int e^{ax}\sin nx\, dx = e^{ax}(a\sin nx - n\cos nx);$$

hence $\int e^{ax}\sin nx\, dx = \dfrac{e^{ax}(a\sin nx - n\cos nx)}{a^2 + n^2}.$

Substituting this in (1) and transposing, gives

$$\frac{a}{n}\int e^{ax}\cos nx\, dx = \frac{e^{ax}(an\sin nx + a^2\cos nx)}{(a^2 + n^2)n};$$

hence $\int e^{ax}\cos nx\, dx = \dfrac{e^{ax}(n\sin nx + a\cos nx)}{a^2 + n^2}.$

EXAMPLES.

1. $\displaystyle\int\frac{dx}{4-5\sin x}=\frac{1}{3}\log\frac{\tan\frac{x}{2}-2}{2\tan\frac{x}{2}-1}.$

2. $\displaystyle\int\frac{dx}{5+4\sin 2x}=\frac{1}{3}\tan^{-1}\frac{5\tan x+4}{3}.$

3. $\displaystyle\int\frac{dx}{3+5\cos x}=\frac{1}{4}\log\frac{\tan\frac{x}{2}+2}{\tan\frac{x}{2}-2}.$

4. $\displaystyle\int\frac{dx}{5-3\cos x}=\frac{1}{2}\tan^{-1}\left(2\tan\frac{x}{2}\right).$

5. $\displaystyle\int\frac{dx}{5-4\cos 2x}=\frac{1}{3}\tan^{-1}(3\tan x).$

6. $\displaystyle\int e^{\frac{x}{2}}\cos\frac{x}{2}dx=e^{\frac{x}{2}}\left(\sin\frac{x}{2}+\cos\frac{x}{2}\right).$

7. $\displaystyle\int\frac{\sin x\,dx}{e^x}=-\frac{\sin x+\cos x}{2e^x}.$

8. $\displaystyle\int e^x\sin^2 x\,dx=\frac{e^x}{2}\left(1-\frac{2\sin 2x+\cos 2x}{5}\right).$

9. $\displaystyle\int e^x\sin 2x\sin x\,dx=\frac{e^x}{4}\left(\sin x+\cos x-\frac{3\sin 3x+\cos 3x}{5}\right).$

10. $\displaystyle\int e^{ax}(\sin ax+\cos ax)\,dx=\frac{e^{ax}\sin ax}{a}.$

11. $\displaystyle\int e^{3x}(\sin 2x-\cos 2x)\,dx=\frac{e^{3x}}{13}(\sin 2x-5\cos 2x).$

CHAPTER VI.

INTEGRALS FOR REFERENCE.

36. We give for reference a list of some of the integrals of the preceding chapters.

1. $\int x^n\, dx = \dfrac{x^{n+1}}{n+1}.$

2. $\int \dfrac{dx}{x} = \log x.$

3. $\int \dfrac{dx}{x^2 + a^2} = \dfrac{1}{a} \tan^{-1} \dfrac{x}{a}.$

4. $\int \dfrac{dx}{x^2 - a^2} = \dfrac{1}{2a} \log \dfrac{x-a}{x+a}.$

Exponential Integrals.

5. $\int a^x\, dx = \dfrac{a^x}{\log a}.$

6. $\int e^x\, dx = e^x.$

Trigonometric Integrals.

7. $\int \sin x\, dx = -\cos x.$

8. $\int \cos x\, dx = \sin x.$

9. $\int \tan x\, dx = \log \sec x.$

10. $\int \cot x\, dx = \log \sin x.$

11. $\displaystyle\int \sec x\,dx = \log(\sec x + \tan x)$
$\displaystyle\qquad\qquad = \log\tan\left(\frac{\pi}{4}+\frac{x}{2}\right).$

12. $\displaystyle\int \csc x\,dx = \log(\csc x - \cot x)$
$\displaystyle\qquad\qquad = \log\tan\frac{x}{2}.$

13. $\displaystyle\int \sec^2 x\,dx = \tan x.$

14. $\displaystyle\int \csc^2 x\,dx = -\cot x.$

15. $\displaystyle\int \sec x\tan x\,dx = \sec x.$

16. $\displaystyle\int \csc x\cot x\,dx = -\csc x.$

17. $\displaystyle\int \sin^2 x\,dx = \frac{x}{2}-\frac{1}{4}\sin 2x.$

18. $\displaystyle\int \cos^2 x\,dx = \frac{x}{2}+\frac{1}{4}\sin 2x.$

INTEGRALS CONTAINING $\sqrt{a^2-x^2}$.

19. $\displaystyle\int \frac{dx}{\sqrt{a^2-x^2}} = \sin^{-1}\frac{x}{a}.$

20. $\displaystyle\int \frac{x^2\,dx}{\sqrt{a^2-x^2}} = -\frac{x}{2}\sqrt{a^2-x^2}+\frac{a^2}{2}\sin^{-1}\frac{x}{a}.$

21. $\displaystyle\int \frac{dx}{x\sqrt{a^2-x^2}} = \frac{1}{a}\log\frac{x}{a+\sqrt{a^2-x^2}}.$

22. $\displaystyle\int \frac{dx}{x^2\sqrt{a^2-x^2}} = -\frac{\sqrt{a^2-x^2}}{a^2 x}.$

23. $\displaystyle\int \frac{dx}{x^3\sqrt{a^2-x^2}} = -\frac{\sqrt{a^2-x^2}}{2a^2 x^2}+\frac{1}{2a^3}\log\frac{x}{a+\sqrt{a^2-x^2}}.$

24. $\int \sqrt{a^2-x^2}\,dx = \frac{x}{2}\sqrt{a^2-x^2} + \frac{a^2}{2}\sin^{-1}\frac{x}{a}.$

25. $\int x^2\sqrt{a^2-x^2}\,dx = \frac{x}{8}(2x^2-a^2)\sqrt{a^2-x^2} + \frac{a^4}{8}\sin^{-1}\frac{x}{a}.$

26. $\int \frac{dx}{(a^2-x^2)^{\frac{3}{2}}} = \frac{x}{a^2\sqrt{a^2-x^2}}.$

27. $\int (a^2-x^2)^{\frac{3}{2}}\,dx = \frac{x}{8}(5a^2-2x^2)\sqrt{a^2-x^2} + \frac{3a^4}{8}\sin^{-1}\frac{x}{a}.$

Integrals containing $\sqrt{x^2+a^2}$.

28. $\int \frac{dx}{\sqrt{x^2+a^2}} = \log(x+\sqrt{x^2+a^2}).$

29. $\int \frac{x^2\,dx}{\sqrt{x^2+a^2}} = \frac{x}{2}\sqrt{x^2+a^2} - \frac{a^2}{2}\log(x+\sqrt{x^2+a^2}).$

30. $\int \frac{dx}{x\sqrt{x^2+a^2}} = \frac{1}{a}\log\frac{x}{a+\sqrt{x^2+a^2}}.$

31. $\int \frac{dx}{x^2\sqrt{x^2+a^2}} = -\frac{\sqrt{x^2+a^2}}{a^2 x}.$

32. $\int \frac{dx}{x^3\sqrt{x^2+a^2}} = -\frac{\sqrt{x^2+a^2}}{2a^2x^2} + \frac{1}{2a^3}\log\frac{a+\sqrt{x^2+a^2}}{x}.$

33. $\int \sqrt{x^2+a^2}\,dx = \frac{x}{2}\sqrt{x^2+a^2} + \frac{a^2}{2}\log(x+\sqrt{x^2+a^2}).$

34. $\int x^2\sqrt{x^2+a^2}\,dx = \frac{x}{8}(2x^2+a^2)\sqrt{x^2+a^2} - \frac{a^4}{8}\log(x+\sqrt{x^2+a^2}).$

35. $\int \frac{dx}{(x^2+a^2)^{\frac{3}{2}}} = \frac{x}{a^2\sqrt{x^2+a^2}}.$

36. $\int (x^2+a^2)^{\frac{3}{2}}\,dx = \frac{x}{8}(2x^2+5a^2)\sqrt{x^2+a^2} + \frac{3a^4}{8}\log(x+\sqrt{x^2+a^2}).$

INTEGRALS CONTAINING $\sqrt{x^2-a^2}$.

37. $\displaystyle\int \frac{dx}{\sqrt{x^2-a^2}} = \log(x+\sqrt{x^2-a^2})$.

38. $\displaystyle\int \frac{x^2 dx}{\sqrt{x^2-a^2}} = \frac{x}{2}\sqrt{x^2-a^2} + \frac{a^2}{2}\log(x+\sqrt{x^2-a^2})$.

39. $\displaystyle\int \frac{dx}{x\sqrt{x^2-a^2}} = \frac{1}{a}\sec^{-1}\frac{x}{a}$.

40. $\displaystyle\int \frac{dx}{x^2\sqrt{x^2-a^2}} = \frac{\sqrt{x^2-a^2}}{a^2 x}$.

41. $\displaystyle\int \frac{dx}{x^3\sqrt{x^2-a^2}} = \frac{\sqrt{x^2-a^2}}{2a^2 x^2} + \frac{1}{2a^3}\sec^{-1}\frac{x}{a}$.

42. $\displaystyle\int \sqrt{x^2-a^2}\, dx = \frac{x}{2}\sqrt{x^2-a^2} - \frac{a^2}{2}\log(x+\sqrt{x^2-a^2})$.

43. $\displaystyle\int x^2\sqrt{x^2-a^2}\, dx = \frac{x}{8}(2x^2-a^2)\sqrt{x^2-a^2} - \frac{a^4}{8}\log(x+\sqrt{x^2-a^2})$.

44. $\displaystyle\int \frac{dx}{(x^2-a^2)^{\frac{3}{2}}} = -\frac{x}{a^2\sqrt{x^2-a^2}}$.

45. $\displaystyle\int (x^2-a^2)^{\frac{3}{2}}\, dx = \frac{x}{8}(2x^2-5a^2)\sqrt{x^2-a^2} + \frac{3a^4}{8}\log(x+\sqrt{x^2-a^2})$.

INTEGRALS CONTAINING $\sqrt{2ax-x^2}$.

46. $\displaystyle\int \frac{dx}{\sqrt{2ax-x^2}} = \operatorname{vers}^{-1}\frac{x}{a}$.

47. $\displaystyle\int \frac{x\, dx}{\sqrt{2ax-x^2}} = -\sqrt{2ax-x^2} + a\operatorname{vers}^{-1}\frac{x}{a}$.

48. $\displaystyle\int \frac{dx}{x\sqrt{2ax-x^2}} = -\frac{\sqrt{2ax-x^2}}{ax}$.

49. $\displaystyle\int \sqrt{2ax-x^2}\, dx = \frac{x-a}{2}\sqrt{2ax-x^2} + \frac{a^2}{2}\operatorname{vers}^{-1}\frac{x}{a}$.

50. $\int x\sqrt{2ax-x^2}\,dx = -\dfrac{3a^2+ax-2x^2}{6}\sqrt{2ax-x^2}+\dfrac{a^3}{2}\operatorname{vers}^{-1}\dfrac{x}{a}.$

51. $\int \dfrac{\sqrt{2ax-x^2}\,dx}{x} = \sqrt{2ax-x^2}+a\operatorname{vers}^{-1}\dfrac{x}{a}.$

52. $\int \dfrac{\sqrt{2ax-x^2}\,dx}{x^3} = -\dfrac{(2ax-x^2)^{\frac{3}{2}}}{3ax^3}.$

53. $\int \dfrac{dx}{(2ax-x^2)^{\frac{3}{2}}} = \dfrac{x-a}{a^2\sqrt{2ax-x^2}}.$

54. $\int \dfrac{x\,dx}{(2ax-x^2)^{\frac{3}{2}}} = \dfrac{x}{a\sqrt{2ax-x^2}}.$

INTEGRALS CONTAINING $\pm ax^2+bx+c$.

55. $\int \dfrac{dx}{ax^2+bx+c} = \dfrac{2}{\sqrt{4ac-b^2}}\tan^{-1}\dfrac{2ax+b}{\sqrt{4ac-b^2}},$

56. $\quad$ or $\quad = \dfrac{1}{\sqrt{b^2-4ac}}\log\dfrac{2ax+b-\sqrt{b^2-4ac}}{2ax+b+\sqrt{b^2-4ac}}.$

57. $\int \dfrac{dx}{\sqrt{ax^2+bx+c}} = \dfrac{1}{\sqrt{a}}\log(2ax+b+2\sqrt{a}\sqrt{ax^2+bx+c}).$

58. $\int \sqrt{ax^2+bx+c}\,dx = \dfrac{2ax+b}{4a}\sqrt{ax^2+bx+c}$
$\qquad -\dfrac{b^2-4ac}{8a^{\frac{3}{2}}}\log(2ax+b+2\sqrt{a}\sqrt{ax^2+bx+c}).$

59. $\int \dfrac{dx}{\sqrt{-ax^2+bx+c}} = \dfrac{1}{\sqrt{a}}\sin^{-1}\dfrac{2ax-b}{\sqrt{b^2+4ac}}.$

60. $\int \sqrt{-ax^2+bx+c}\,dx$
$\quad = \dfrac{2ax-b}{4a}\sqrt{-ax^2+bx+c}+\dfrac{b^2+4ac}{8a^{\frac{3}{2}}}\sin^{-1}\dfrac{2ax-b}{\sqrt{b^2+4ac}}.$

Other Integrals.

61. $\int \sqrt{\dfrac{a+x}{b+x}}\,dx$

$= \sqrt{(a+x)(b+x)} + (a-b)\log(\sqrt{a+x}+\sqrt{b+x}).$

62. $\int \sqrt{\dfrac{a-x}{b+x}}\,dx = \sqrt{(a-x)(b+x)} + (a+b)\sin^{-1}\sqrt{\dfrac{x+b}{a+b}}.$

CHAPTER VII.

INTEGRATION AS A SUMMATION. DEFINITE INTEGRALS.

37. The process of integration may be regarded as the summation of an infinite series of infinitely small terms. As an illustration, consider the following problem.

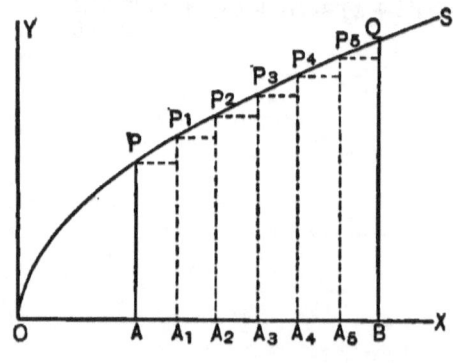

38. *To find the area $PABQ$ included between a given curve OS, the axis of X, and the ordinates AP and BQ.*

Let $y = x^{\frac{1}{2}}$ be the equation of the given curve.
Let $OA = a$, $OB = b$.
Suppose AB divided into n equal parts (in the figure, $n = 6$), and let Δx denote one of the equal parts, as AA_1, A_1A_2, ⋯.

Then $\qquad AB = b - a = n\Delta x$.

At $A_1, A_2, \cdots$, draw the ordinates A_1P_1, A_2P_2, ⋯, and complete the rectangles PA_1, P_1A_2, ⋯.

From the equation of the curve, $y = x^{\frac{1}{2}}$,

$PA = a^{\frac{1}{2}}$, $P_1A_1 = (a + \Delta x)^{\frac{1}{2}}$, $P_2A_2 = (a + 2\Delta x)^{\frac{1}{2}}$, ⋯ $QB = b^{\frac{1}{2}}$.

Area of rectangle $PA_1 = PA \times AA_1 = a^{\frac{1}{2}}\Delta x$.

Area of rectangle $P_1A_2 = P_1A_1 \times A_1A_2 = (a + \Delta x)^{\frac{1}{2}}\Delta x$.

Area of rectangle $P_2A_3 = P_2A_2 \times A_2A_3 = (a + 2\Delta x)^{\frac{1}{2}}\Delta x$.

⋯ ⋯ ⋯ ⋯ ⋯ ⋯

INTEGRATION AS A SUMMATION. 237

The sum of all the n rectangles is

$$a^{\frac{1}{2}}\Delta x + (a + \Delta x)^{\frac{1}{2}} \Delta x + (a + 2 \Delta x)^{\frac{1}{2}} \Delta x + \cdots + (b - \Delta x)^{\frac{1}{2}} \Delta x,$$

which may be represented by $\sum_{a}^{b} x^{\frac{1}{2}} \Delta x$,

where $x^{\frac{1}{2}}\Delta x$ represents each term of the series, x taking in succession the values a, $a + \Delta x$, $a + 2 \Delta x$, $\cdots$, $b - \Delta x$.

It is evident that the area $PABQ$ is the limit of the sum of the rectangles, as n increases, and Δx decreases.

When Δx in the preceding series is changed into the infinitesimal dx, the symbol $\sum$ is replaced by $\int$, an abbreviation of the word "sum."

Thus

$$\int_{a}^{b} x^{\frac{1}{2}} dx = a^{\frac{1}{2}} dx + (a + dx)^{\frac{1}{2}} dx + (a + 2 dx)^{\frac{1}{2}} dx + \cdots + (b - dx)^{\frac{1}{2}} dx$$
$$= \text{area } PABQ. \quad \ldots \ldots \ldots \ldots \quad (1)$$

The expression $\int_{a}^{b} x^{\frac{1}{2}} dx$, as defined by (1), denotes the sum of an infinite number of terms, each of which is represented by $x^{\frac{1}{2}} dx$, x taking in succession the values a, $a + dx$, $a + 2 dx$, $\cdots$, $b - dx$.

Or the definition may be more precisely expressed by

$$\int_{a}^{b} x^{\frac{1}{2}} dx = \text{Limit of } \sum_{a}^{b} x^{\frac{1}{2}} \Delta x, \quad \text{as } \Delta x \text{ approaches zero.}$$

It is to be noticed that a new meaning is thus given to the symbol $\int$, which has been defined heretofore as the inverse of differentiation. It will now be shown that the two definitions are consistent.

39. *Value of* $\int_{a}^{b} x^{\frac{1}{2}} dx$. To find the area $PABQ$ we must find the sum of the series (1), Art. 38, that is, the value of $\int_{a}^{b} x^{\frac{1}{2}} dx$.

Now $\qquad x^{\frac{1}{2}} dx = d\left(\dfrac{2}{3} x^{\frac{3}{2}}\right).$

But the differential of a function of x is the increment of that function when x receives the increment dx.

That is, $d\left(\dfrac{2}{3}x^{\frac{3}{2}}\right) = \dfrac{2}{3}(x+dx)^{\frac{3}{2}} - \dfrac{2}{3}x^{\frac{3}{2}}.$

Hence, $x^{\frac{1}{2}}dx = \dfrac{2}{3}(x+dx)^{\frac{3}{2}} - \dfrac{2}{3}x^{\frac{3}{2}}.$

Substituting for x, a, $a+dx$, $a+2\,dx$, $\cdots$, $b-dx$,
we have $a^{\frac{1}{2}}dx = \dfrac{2}{3}(a+dx)^{\frac{3}{2}} - \dfrac{2}{3}a^{\frac{3}{2}},$

$$(a+dx)^{\frac{1}{2}}dx = \dfrac{2}{3}(a+2\,dx)^{\frac{3}{2}} - \dfrac{2}{3}(a+dx)^{\frac{3}{2}},$$

$$(a+2\,dx)^{\frac{1}{2}}dx = \dfrac{2}{3}(a+3\,dx)^{\frac{3}{2}} - \dfrac{2}{3}(a+2\,dx)^{\frac{3}{2}},$$

$$\cdots\cdots\cdots\cdots\cdots\cdots\cdots$$

$$(b-dx)^{\frac{1}{2}}dx = \dfrac{2}{3}b^{\frac{3}{2}} - \dfrac{2}{3}(b-dx)^{\frac{3}{2}}.$$

Adding and cancelling terms in second member, we have

$$a^{\frac{1}{2}}dx + (a+dx)^{\frac{1}{2}}dx + (a+2\,dx)^{\frac{1}{2}} + \cdots + (b-dx)^{\frac{1}{2}}dx = \dfrac{2}{3}b^{\frac{3}{2}} - \dfrac{2}{3}a^{\frac{3}{2}}.$$

That is, as x varies from a to b, the sum of the successive increments of the function $\dfrac{2}{3}x^{\frac{3}{2}}$ is equal to its entire increment.

Thus $\displaystyle\int_a^b x^{\frac{1}{2}}dx = \dfrac{2}{3}b^{\frac{3}{2}} - \dfrac{2}{3}a^{\frac{3}{2}} = \text{area } PABQ.$

We have thus shown that the sum of the infinite series represented by $\displaystyle\int_a^b x^{\frac{1}{2}}dx$ is found by substituting for x, b and a in $\dfrac{2}{3}x^{\frac{3}{2}}$, the integral of $x^{\frac{1}{2}}dx$, and subtracting the latter result from the former.

The expression $\displaystyle\int_a^b x^{\frac{1}{2}}dx$ is called a *definite integral*, and the process of evaluating it is called *integrating between limits*, the initial value a of the variable being the *inferior limit*, and the final value b the *superior limit*.

In contradistinction $\dfrac{2}{3}x^{\frac{3}{2}}$ is called the *indefinite integral* of $x^{\frac{1}{2}}dx$.

40. The relation of the terms of the series $\int_a^b x^{\frac{1}{2}}dx$ to the integral $\frac{2}{3}x^{\frac{3}{2}}$ may be made clearer to the student by considering the following series of numbers:

$$\begin{array}{cc} 1 & \\ & 3 \\ 4 & \\ & 5 \\ 9 & \\ & 7 \\ 16 & \\ & 9 \\ 25 & \\ & 11 \\ 36 & \end{array}$$

The numbers in the second column are the differences between consecutive numbers in the first, and it is evident that the sum of the second column of numbers is the difference between the first and last, in the first column. That is,

$$3 + 5 + 7 + 9 + 11 = 36 - 1.$$

The terms of $\int_a^b x^{\frac{1}{2}}dx$ may be similarly arranged, as follows:

$\frac{2}{3}a^{\frac{3}{2}},$

$\qquad\qquad a^{\frac{1}{2}}dx,$

$\frac{2}{3}(a+dx)^{\frac{3}{2}},$

$\qquad\qquad (a+dx)^{\frac{1}{2}}dx,$

$\frac{2}{3}(a+2\,dx)^{\frac{3}{2}},$

$\qquad\qquad (a+2\,dx)^{\frac{1}{2}}dx,$

$\frac{2}{3}(a+3\,dx)^{\frac{3}{2}},$

$\qquad \cdots \qquad \cdots \qquad\qquad \cdots \qquad \cdots$

$\frac{2}{3}(b-dx)^{\frac{3}{2}},$

$\qquad\qquad (b-dx)^{\frac{1}{2}}dx.$

$\frac{2}{3}b^{\frac{3}{2}}.$

Since $x^{\frac{1}{2}}dx$ is the differential of $\frac{2}{3}x^{\frac{3}{2}}$, the terms in the second column are the infinitesimal differences between the consecutive terms in the first, and therefore

$$a^{\frac{1}{2}}dx + (a+dx)^{\frac{1}{2}}dx + (a+2\,dx)^{\frac{1}{2}} + \cdots + (b-dx)^{\frac{1}{2}}dx = \frac{2}{3}b^{\frac{3}{2}} - \frac{2}{3}a^{\frac{3}{2}};$$

that is, $\qquad \int_a^b x^{\frac{1}{2}}dx = \frac{2}{3}b^{\frac{3}{2}} - \frac{2}{3}a^{\frac{3}{2}}.$

41. *General Definition of a Definite Integral.*

In general, if $\phi(x)$ denote any given function of x, which is finite and continuous from $x=a$ to $x=b$, $\int_a^b \phi(x)dx$ is the definite integral representing the sum of an infinite series of terms, obtained from $\phi(x)dx$, by supposing x to vary from a to b.

If $\qquad \int \phi(x)dx = \psi(x)$, the indefinite integral,

then $\qquad \int_a^b \phi(x)dx = \psi(b) - \psi(a).$

This may be illustrated by an area as in Art. 38, by supposing $y = \phi(x)$ to be the equation of the curve OS, and the proof of Art. 39 may be similarly modified, by substituting $\phi(x)$ for $x^{\frac{1}{2}}$, and $\psi(x)$ for $\frac{2}{3}x^{\frac{3}{2}}$.

42. We add in this article the proof of the relation between the definite and indefinite integrals, expressed in the form of limits instead of infinitesimals as in Art. 39.

We shall use the expression "$\text{Limit}_{\Delta x \doteq 0}$" to denote the words "The limit, as Δx approaches zero, of."

Given $\qquad \phi(x) = \frac{d}{dx}\psi(x), \qquad$ and

$$\sum_a^b \phi(x)\Delta x = \phi(a)\Delta x + \phi(a + \Delta x)\Delta x + \phi(a + 2\,\Delta x)\Delta x + \cdots + \phi(b - \Delta x)\Delta x,$$

the function $\phi(x)$ being finite and continuous from $x = a$ to $x = b$; to prove that

$$\text{Limit}_{\Delta x = 0} \sum_{a}^{b} \phi(x)\Delta x = \psi(b) - \psi(a).$$

From the definition of $\dfrac{d}{dx}\psi(x)$, in Art. 10, Dif. Cal.,

$$\phi(x) = \frac{d}{dx}\psi(x) = \text{Limit}_{\Delta x = 0}\frac{\psi(x+\Delta x) - \psi(x)}{\Delta x}.$$

Hence $\dfrac{\psi(x+\Delta x) - \psi(x)}{\Delta x} = \phi(x) + \epsilon,$

where ϵ is a quantity that vanishes with Δx. Hence,

$$\psi(x+\Delta x) - \psi(x) = \phi(x)\Delta x + \epsilon \Delta x.$$

Substituting in this equation for x,

$$a, \quad a + \Delta x, \quad a + 2\Delta x, \quad \cdots \quad b - \Delta x,$$

we have

$$\psi(a+\Delta x) - \psi(a) = \phi(a)\Delta x + \epsilon_1 \Delta x,$$
$$\psi(a+2\Delta x) - \psi(a+\Delta x) = \phi(a+\Delta x)\Delta x + \epsilon_2 \Delta x,$$
$$\psi(a+3\Delta x) - \psi(a+2\Delta x) = \phi(a+2\Delta x)\Delta x + \epsilon_3 \Delta x,$$
$$\cdots \quad \cdots \quad \cdots \quad \cdots \quad \cdots \quad \cdots$$
$$\psi(b) - \psi(b-\Delta x) = \phi(b-\Delta x)\Delta x + \epsilon_n \Delta x.$$

Adding and cancelling terms in first member, we find

$$\psi(b) - \psi(a) = \sum_{a}^{b}\phi(x)\Delta x + \sum_{a}^{b}\epsilon \Delta x. \quad \cdots \quad (1)$$

Now if ϵ_k is the greatest of the quantities $\epsilon_1, \epsilon_2, \cdots \epsilon_n,$ it follows that

$$\sum_{a}^{b}\epsilon \Delta x < \epsilon_k \sum_{a}^{b}\Delta x;$$

that is, $\sum_{a}^{b}\epsilon \Delta x < \epsilon_k(b-a).$

Hence $\sum_a^b \epsilon \Delta x$ vanishes with ϵ, that is, with Δx.

Taking the limit of (1), we have

$$\psi(b) - \psi(a) = \text{Limit}_{\Delta x = 0} \sum_a^b \phi(x) \Delta x = \int_a^b \phi(x) dx.$$

43. It is to be noticed that the arbitrary constant c, in the indefinite integral, disappears from the definite integral.

Thus, if in evaluating $\int_a^b x^3 dx$, we call the indefinite integral $\frac{x^4}{4} + c$, we have

$$\int_a^b x^3 dx = \frac{b^4}{4} + c - \left(\frac{a^4}{4} + c\right) = \frac{b^4}{4} - \frac{a^4}{4}, \text{ as before.}$$

Or if in evaluating $\int_a^b \phi(x) dx$, we call the indefinite integral $\psi(x) + c$, we have

$$\int_a^b \phi(x) dx = \psi(b) + c - [\psi(a) + c] = \psi(b) - \psi(a),$$

as before.

EXAMPLES.

Evaluate the following definite integrals:

1. $\int_1^4 x^2 dx = \frac{x^3}{3}\Big|_1^4 = \frac{64}{3} - \frac{1}{3} = 21.$

2. $\int_1^e \frac{dx}{x} = \log x \Big|_1^e = \log e - \log 1 = 1.$

3. $\int_0^{\frac{\pi}{2}} \sin x \, dx = -\cos x \Big|_0^{\frac{\pi}{2}} = 0 - (-1) = 1.$

4. $\int_0^b (b^2 x - x^3) dx = \frac{b^4}{4}.$

5. $\int_1^4 \frac{dx}{x^{\frac{3}{2}}} = 1.$

6. $\int_2^3 \dfrac{x\,dx}{1+x^2} = \dfrac{\log 2}{2}$.

7. $\int_0^\infty \dfrac{8a^3\,dx}{x^2+4a^2} = 2\pi a^2$.

8. $\int_0^{\frac{\pi}{4}} \sec^4\theta\,d\theta = \dfrac{4}{3}$.

9. $\int_1^e x\log x\,dx = \dfrac{e^2+1}{4}$.

10. $\int_{\frac{\pi}{6}}^{\frac{\pi}{4}} \dfrac{dx}{\cos x} = \log\left(\dfrac{1+\sqrt{2}}{\sqrt{3}}\right)$.

11. $\int_1^\infty \dfrac{dx}{x^2 - 2x\cos a + 1} = \dfrac{\pi - a}{2\sin a}$.

12. $\int_0^\infty \dfrac{dx}{(a^2+x^2)(b^2+x^2)} = \dfrac{\pi}{2ab(a+b)}$.

13. $\int_0^\infty e^{-nx}\sin nx\,dx = \dfrac{1}{2n}$.

14. $\int_0^{\frac{\pi}{2}} \dfrac{dx}{2+\cos x} = \dfrac{\pi}{3\sqrt{3}}$.

Derive the following by (5) and (7), Art. 31:

15. If n is even,

$$\int_0^{\frac{\pi}{2}} \sin^n x\,dx = \int_0^{\frac{\pi}{2}} \cos^n x\,dx = \dfrac{1\cdot 3\cdot 5\cdots (n-1)}{2\cdot 4\cdot 6\cdots n}\dfrac{\pi}{2}.$$

16. If n is odd,

$$\int_0^{\frac{\pi}{2}} \sin^n x\,dx = \int_0^{\frac{\pi}{2}} \cos^n x\,dx = \dfrac{2\cdot 4\cdot 6\cdots (n-1)}{3\cdot 5\cdot 7\cdots n}.$$

43½. Change of Limits. When a new variable is used in obtaining the indefinite integral, we may avoid the restoration of the original variable, by changing the limits to correspond with the new variable.

For example, to evaluate

$$\int_0^4 \frac{dx}{1+\sqrt{x}}, \quad \text{assume } \sqrt{x} = z.$$

Then we have $\quad \dfrac{dx}{1+\sqrt{x}} = \dfrac{2z\,dz}{1+z}.$

Now when $x = 4$, $z = 2$; and when $x = 0$, $z = 0$.

Hence $\quad \displaystyle\int_0^4 \frac{dx}{1+\sqrt{x}} = \int_0^2 \frac{2z\,dz}{1+z} = 2\left[z - \log(1+z)\right]_0^2$

$$= 4 - 2\log 3.$$

EXAMPLES.

1. $\displaystyle\int_{\frac{1}{4}}^1 \frac{(x - x^3)^{\frac{1}{3}}\,dx}{x^4} = 6.$ $\qquad$ Assume $x = \dfrac{1}{z}$.

2. $\displaystyle\int_3^{29} \frac{(x-2)^{\frac{1}{3}}\,dx}{(x-2)^{\frac{2}{3}} + 3} = 8 + \dfrac{3\sqrt{3}}{2}\pi.$ $\quad$ Assume $x - 2 = z^3$.

3. $\displaystyle\int_0^{\log 5} \frac{e^x\sqrt{e^x - 1}}{e^x + 3}\,dx = 4 - \pi.$ $\qquad$ Assume $e^x - 1 = z^2$.

4. $\displaystyle\int_0^\infty \frac{dx}{\sqrt{e^{2x} + \tan^2 a}} = \dfrac{1}{\tan a}\log(\sec a + \tan a).$

$\qquad\qquad\qquad\qquad\qquad\qquad\qquad$ Assume $e^{2x} + \tan^2 a = z^2$.

5. $\displaystyle\int_0^{\frac{\pi}{4}} \frac{(\sin\theta + \cos\theta)\,d\theta}{3 + \sin 2\theta} = \dfrac{\log 3}{4}.$ $\quad$ Assume $\sin\theta - \cos\theta = x$.

6. $\displaystyle\int_1^{2+\sqrt{5}} \frac{(x^2 + 1)\,dx}{x\sqrt{x^4 + 7x^2 + 1}} = \log 3.$ $\quad$ Assume $x - \dfrac{1}{x} = z$.

CHAPTER VIII.

APPLICATION OF INTEGRATION TO PLANE CURVES. APPLICATION TO CERTAIN VOLUMES.

44. *Areas of Curves. Rectangular Co-ordinates.* The simplest application of integration to curves is in determining the areas defined by them. We have already used this problem in Arts. 38, 39, as an illustration of a definite integral. We shall now consider it in a more general form, and derive the expression for the area included by *any* curve, in rectangular co-ordinates.

45. *To find the area between a given curve PQ, the axis of X, and two given ordinates, AP and BQ.*

Let $OA = a$, $OB = b$.
Let x and y be the co-ordinates of any point P_2 of the curve; then

$$x + \Delta x, \quad y + \Delta y,$$

will be the co-ordinates of P_3.

The area of the rectangle $P_2 A_2 A_3$ is

$$P_2 A_2 \times A_2 A_3 = y \Delta x.$$

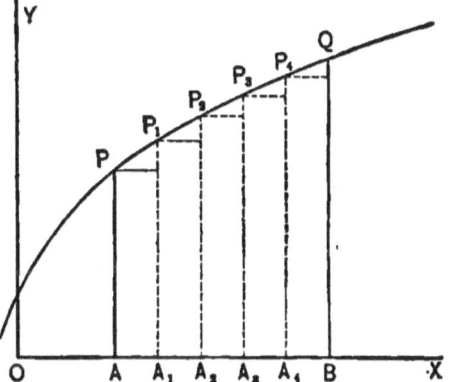

The sum of all the rectangles PAA_1, $P_1 A_1 A_2$, $P_2 A_2 A_3$, ..., may be represented by $\sum_a^b y \Delta x.$

The required area $PQBA$ is the limit of the sum of the rectangles, as Δx is indefinitely diminished. That is

$$A = \int_a^b y\, dx.$$

We may also regard the required area as generated, by the ordinate AP moving from left to right, and varying in length according to the equation of the given curve. Regarding y as constant while moving the distance dx, it generates the rectangle $y\,dx$. Then the general formula for the required area is

$$A = \int_a^b y\,dx, \quad \text{as before;}$$

the inferior limit $a = OA$, denoting the initial position of the moving ordinate, and the superior limit $b = OB$, its final position.

Similarly the area between the given curve, the axis of Y, and two given abscissas, is

$$A = \int x\,dy,$$

the limits of integration being the limiting values of y.

EXAMPLES.

1. Find the area between the parabola $y^2 = 4ax$ and the axis of X, from the origin to the ordinate at the point (h, k).

   Here $\quad A = \int_0^h y\,dx = \int_0^h 2a^{\frac{1}{2}}x^{\frac{1}{2}}\,dx = \dfrac{4a^{\frac{1}{2}}x^{\frac{3}{2}}}{3}\Big|_0^h = \dfrac{4a^{\frac{1}{2}}h^{\frac{3}{2}}}{3}.$

   Since $\quad k^2 = 4ah, \quad k = 2a^{\frac{1}{2}}h^{\frac{1}{2}}.$

   $\therefore\ A = \dfrac{2}{3}hk$, two-thirds the circumscribed rectangle.

2. Find the entire area of the ellipse $\dfrac{x^2}{a^2} + \dfrac{y^2}{b^2} = 1$. *Ans.* πab.

3. Show that the area of a sector of the equilateral hyperbola $x^2 - y^2 = a^2$, included between the axis of X and a diameter through the point (x, y) of the curve, is

   $$\dfrac{a^2}{2}\log\dfrac{x+y}{a}.$$

4. Find the entire area between the witch $y = \dfrac{8a^3}{x^2 + 4a^2}$, and the axis of X.

   Ans. $4\pi a^2$.

AREAS OF CURVES.   247

5. Find the area intercepted between the co-ordinate axes by the parabola $x^{\frac{1}{2}} + y^{\frac{1}{2}} = a^{\frac{1}{2}}$.   Ans. $\dfrac{a^2}{6}$.

6. Find the entire area within the curve $\left(\dfrac{x}{a}\right)^2 + \left(\dfrac{y}{b}\right)^{\frac{3}{2}} = 1$.   Ans. $\dfrac{3}{4}\pi ab$.

7. Find the entire area within the hypocycloid $x^{\frac{2}{3}} + y^{\frac{2}{3}} = a^{\frac{2}{3}}$.   Ans. $\dfrac{3\pi a^2}{8}$.

8. Find the entire area between the cissoid $y^2 = \dfrac{x^3}{2a-x}$, and the line $x = 2a$, its asymptote.   Ans. $3\pi a^2$.

The area between two curves is the sum, or the difference, of the areas between the curves and one of the co-ordinate axes, the limits being determined by the points of intersection.

9. Find the area included between the parabola $x^2 = 4ay$, and the witch $y = \dfrac{8a^3}{x^2 + 4a^2}$.   Ans. $\left(2\pi - \dfrac{4}{3}\right)a^2$.

46. Areas of Curves. Polar Co-ordinates. To find the area POQ, included between a given curve PQ, and two given radii vectores, OP and OQ. Let

$POX = a, \quad QOX = \beta$.

Let r and θ be the co-ordinates of any point P_2 of the curve, then

$r + \Delta r, \quad \theta + \Delta \theta$,

will be the co-ordinates of P_3.

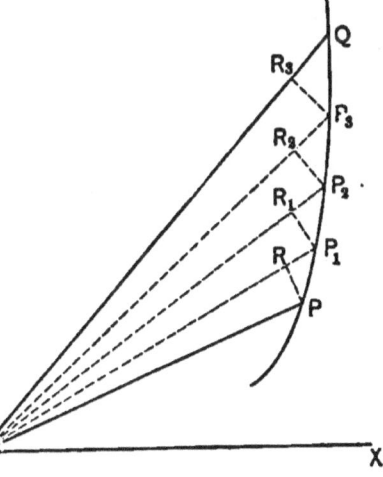

The area of the circular sector P_2OR_2 is

$$\frac{1}{2}OP_2 \times P_2R_2 = \frac{1}{2}r \cdot r\Delta\theta = \frac{1}{2}r^2\Delta\theta.$$

The sum of the sectors POR, P_1OR_1, $P_2OR_2 \cdots$, may be represented by

$$\sum_a^\beta \frac{1}{2} r^2 \Delta\theta.$$

The required area POQ is the limit of the sum of the sectors, as $\Delta\theta$ approaches zero. That is,

$$A = \frac{1}{2}\int_a^\beta r^2 d\theta.$$

47. We may also regard the area POQ as generated, by the radius vector revolving from OP to OQ, and varying in length according to the equation of the given curve.

Regarding r as constant while describing the angle $d\theta$, it generates the sector whose area is $\frac{1}{2}r^2 d\theta$.

Hence $\quad A = \frac{1}{2}\int_a^\beta r^2 d\theta$, as before;

the inferior limit a denoting the initial, and the superior limit β, the final position, of the moving radius vector.

EXAMPLES.

1. Find the area described by the radius vector in one entire revolution of the spiral of Archimedes $r = a\theta$.

Here $A = \dfrac{1}{2}\int_0^{2\pi} r^2 d\theta = \dfrac{1}{2}\int_0^{2\pi} a^2\theta^2 d\theta = \dfrac{a^2}{2}\dfrac{\theta^3}{3}\Big|_0^{2\pi} = \dfrac{4\pi^3 a^2}{3}.$

2. Find the area described by the radius vector in the logarithmic spiral $r = e^{a\theta}$, from $\theta = 0$ to $\theta = \dfrac{\pi}{2}$.

$$\text{Ans. } \frac{1}{4a}(e^{\pi a} - 1).$$

LENGTHS OF CURVES. 249

3. Find the entire area of the circle $r = a\sin\theta$.   Ans. $\dfrac{\pi a^2}{4}$.

4. Find the area of one loop of the curve $r = a\sin 2\theta$.

                                                                    Ans. $\dfrac{\pi a^2}{8}$.

5. Find the entire area of the cardioid $r = a(1 - \cos\theta)$.
     Ans. $\dfrac{3\pi a^2}{2}$, or six times the area of the generating circle.

6. Find the area described by the radius vector in the parabola $r = a\sec^2\dfrac{\theta}{2}$, from $\theta = 0$ to $\theta = \dfrac{\pi}{2}$.   Ans. $\dfrac{4a^2}{3}$.

7. Find the area below OX within the curve $r = a\sin^3\dfrac{\theta}{3}$. p 101

                       Ans. $(10\pi + 27\sqrt{3})\dfrac{a^2}{64}$.

48. Lengths of Curves. Rectangular Co-ordinates. To find the length of the arc PQ between two given points P and Q.

Let $OA = a$, $OB = b$.

Denoting the required length of arc by s, we have

$$ds = \left[1 + \left(\dfrac{dy}{dx}\right)^2\right]^{\frac{1}{2}} dx\,;$$

    (1) Art. 98, Dif. Cal.

therefore

$$s = \int_a^b \left[1 + \left(\dfrac{dy}{dx}\right)^2\right]^{\frac{1}{2}} dx\,;$$

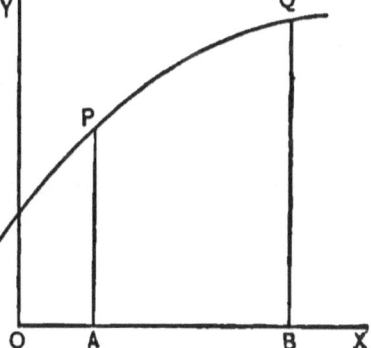

the limits of integration being the limiting values of x.

Or we may evidently use the formula

$$s = \int \left[1 + \left(\dfrac{dx}{dy}\right)^2\right]^{\frac{1}{2}} dy,$$

the limits being the limiting values of y.

EXAMPLES.

1. Find the length of the arc of the parabola $y^2 = 4ax$, from the vertex to the extremity of the latus rectum.

Here
$$\frac{dy}{dx} = \frac{a^{\frac{1}{2}}}{x^{\frac{1}{2}}},$$

therefore $s = \int_0^a \left(1 + \frac{a}{x}\right)^{\frac{1}{2}} dx = \int_0^a \left(\frac{a+x}{x}\right)^{\frac{1}{2}} dx.$

This may be integrated by 9, p. 213, making $b = 0$.

$$\int \left(\frac{a+x}{x}\right)^{\frac{1}{2}} dx = \sqrt{ax + x^2} + a \log(\sqrt{a+x} + \sqrt{x}).$$

$$\int_0^a \left(\frac{a+x}{x}\right)^{\frac{1}{2}} dx = a[\sqrt{2} + \log(1+\sqrt{2})] = 2.29558\,a.$$

2. Find the length of the arc of the semi-cubical parabola $ay^2 = x^3$, from the origin to $x = 5a$. *Ans.* $\dfrac{335\,a}{27}$.

3. Find the length of the arc of the curve $9ay^2 = x(x - 3a)^2$, from $x = 0$ to $x = 3a$. *Ans.* $2a\sqrt{3}$.

4. Find the length of the arc of the catenary $y = \dfrac{a}{2}(e^{\frac{x}{a}} + e^{-\frac{x}{a}})$, from $x = 0$ to the point (x, y).

Ans. $\dfrac{a}{2}(e^{\frac{x}{a}} - e^{-\frac{x}{a}})$.

5. Find the entire length of the arc of the hypocycloid $x^{\frac{2}{3}} + y^{\frac{2}{3}} = a^{\frac{2}{3}}$. *Ans.* $6a$.

49. Lengths of Curves. Polar Co-ordinates. To find the length of the arc PQ between two given points P and Q.

Let $POX = \alpha$, $QOX = \beta$.

LENGTHS OF CURVES.

We have
$$ds = \left[r^2 + \left(\frac{dr}{d\theta}\right)^2\right]^{\frac{1}{2}} d\theta;$$
(3) Art. 98½, Dif. Cal.

therefore
$$s = \int_\alpha^\beta \left[r^2 + \left(\frac{dr}{d\theta}\right)^2\right]^{\frac{1}{2}} d\theta, \quad (1)$$

the limits being the limiting values of θ.

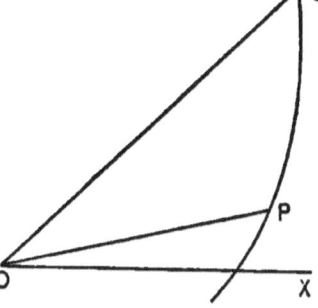

Or we have $ds = \left[1 + r^2\left(\frac{d\theta}{dr}\right)^2\right]^{\frac{1}{2}} dr;$ (2) Art. 98½, Dif. Cal.

therefore
$$s = \int_a^b \left[1 + r^2\left(\frac{d\theta}{dr}\right)^2\right]^{\frac{1}{2}} dr, \quad \ldots \quad (2)$$

the limits being the limiting values of r. That is, $OP = a$, $OQ = b$.

EXAMPLES.

1. Find the length of the arc of the spiral of Archimedes $r = a\theta$, from the pole to the end of the first revolution.

Here $\dfrac{dr}{d\theta} = a.$

$$s = \int_0^{2\pi} (a^2\theta^2 + a^2)^{\frac{1}{2}} d\theta = a \int_0^{2\pi} (1 + \theta^2)^{\frac{1}{2}} d\theta$$
$$= a\left[\frac{\theta\sqrt{1+\theta^2}}{2} + \frac{1}{2}\log(\theta + \sqrt{1+\theta^2})\right]_0^{2\pi}$$
$$= a\left[\pi\sqrt{1+4\pi^2} + \frac{1}{2}\log(2\pi + \sqrt{1+4\pi^2})\right].$$

2. Find the entire length of the cardioid $r = a(1 - \cos\theta)$.
Ans. $8a$.

3. Find the length of the logarithmic spiral $r = e^{a\theta}$, from the pole to the point (r, θ). Use formula (2).
Ans. $\dfrac{r}{a}\sqrt{a^2 + 1}$.

4. Find the entire length of the curve $r = a\sin^3\dfrac{\theta}{3}$.

Ans. $\dfrac{3\pi a}{2}$.

5. The equation of the epicycloid, the radius of the fixed circle being a, and that of the rolling circle $\dfrac{a}{2}$, is

$\sin^2\theta = \dfrac{4(r^2 - a^2)^3}{27\,a^4 r^2}$. Find the length of one loop.

From the above equation

$\dfrac{d\theta}{dr} = \dfrac{2\sqrt{r^2 - a^2}}{r\sqrt{4a^2 - r^2}}$; then use Formula (2). Ans. $6a$.

50. Surfaces of Revolution. Volume. To find the volume generated, by revolving about OX the plane area $APQB$.

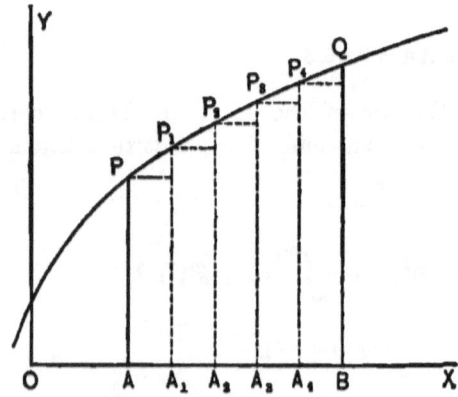

Let $OA = a$, $OB = b$.
Let x and y be the co-ordinates of any point P_2 of the given curve.

It is evident that the rectangle $P_2 A_2 A_3$ will generate a right cylinder, whose volume is $\pi y^2 \Delta x$.

The sum of all these cylinders may be represented by $\pi \sum_a^b y^2 \Delta x$.

The required volume is the limit of the sum of the cylinders, as Δx approaches zero. That is,

$$V = \pi \int_a^b y^2 dx.$$

Or we may regard the required volume as generated by the area of a circle, which moves with its plane always perpendicular to the axis of X, its centre moving along this axis, and its radius being the ordinate of the given curve.

SURFACES OF REVOLUTION. 253

Since y is the radius of this moving circle, its area is πy^2, and regarding y as constant while it moves over the distance dx, we have for the volume of an elementary cylinder,

$$dV = \pi y^2 dx.$$

Hence $\qquad V = \pi \int y^2 dx, \qquad \ldots \ldots \ldots$ (1)

the limits being the limiting values of x.

Similarly, if Y is the axis of revolution,

$$V = \pi \int x^2 dy,$$

the limits being the limiting values of y.

51. *Surfaces of Revolution. Area.* To find the area of the surface generated, by revolving about OX the arc PQ.

In the figure of Art. 50, let P_2P_3 be an element of the given curve.

This will generate the convex surface of the frustum of a right cone. Hence we have by geometry, for an element of the required surface,

$$dS = 2\pi \left(\frac{P_2A_2 + P_3A_3}{2} \right) P_2P_3 = \pi(y + y')ds,$$

where $\qquad y = P_2A_2$ and $y' = P_3A_3$.

But since the limit of y' is y, we have

$$dS = 2\pi y\, ds;$$

hence $\qquad S = 2\pi \int y\, ds.$

By (1) Art. 98, Dif. Cal.,

$$S = 2\pi \int_a^b y \left[1 + \left(\frac{dy}{dx} \right)^2 \right]^{\frac{1}{2}} dx. \quad \ldots \ldots \quad (1)$$

Similarly if OY is the axis of revolution,

$$S = 2\pi \int x\, ds.$$

EXAMPLES.

1. Find the volume and surface of the prolate spheroid obtained, by revolving about X the ellipse $\dfrac{x^2}{a^2}+\dfrac{y^2}{b^2}=1$.

   From (1) Art. 50, we have
   $$\tfrac{1}{2}V = \pi\int_0^a y^2 dx = \pi\int_0^a \dfrac{b^2}{a^2}(a^2-x^2)\,dx = \dfrac{2\pi ab^2}{3}.$$
   $$\therefore\ V = \dfrac{4\pi ab^2}{3}.$$

   From (1) Art. 51,
   $$\tfrac{1}{2}S = 2\pi\int_0^a y\left[1+\left(\dfrac{dy}{dx}\right)^2\right]^{\frac{1}{2}}dx$$
   $$= 2\pi\int_0^a \dfrac{b}{a}\sqrt{a^2-x^2}\left[1+\dfrac{b^2 x^2}{a^2(a^2-x^2)}\right]^{\frac{1}{2}}dx$$
   $$= 2\pi\dfrac{b}{a^2}\int_0^a [a^4-(a^2-b^2)x^2]^{\frac{1}{2}}dx$$
   $$= \pi b\left(b+\dfrac{a^2}{\sqrt{a^2-b^2}}\sin^{-1}\dfrac{\sqrt{a^2-b^2}}{a}\right).$$
   $$\therefore\ S = 2\pi b\left(b+\dfrac{a^2}{\sqrt{a^2-b^2}}\cos^{-1}\dfrac{b}{a}\right).$$

2. Find the volume and surface generated, by revolving about X the parabola $y^2=4ax$, from the origin to $x=a$.

   Ans. $2\pi a^3$ and $\dfrac{8(\sqrt{8}-1)}{3}\pi a^2$.

3. Find the volume and convex surface of the right cone generated, by revolving about X the line joining the origin and the point (a, b). Ans. $\dfrac{\pi a b^2}{3}$ and $\pi b\sqrt{a^2+b^2}$.

4. Find the entire volume and surface generated, by revolving about X the hypocycloid $x^{\frac{2}{3}}+y^{\frac{2}{3}}=a^{\frac{2}{3}}$.

   Ans. $\dfrac{32\pi a^3}{105}$ and $\dfrac{12\pi a^2}{5}$.

OTHER VOLUMES.

5. Find the entire volume generated by revolving the witch $y = \dfrac{8a^3}{x^2 + 4a^2}$ about X, its asymptote. *Ans.* $4\pi^2 a^3$.

6. Find the volume generated by revolving about X, the part of the parabola $x^{\frac{1}{2}} + y^{\frac{1}{2}} = a^{\frac{1}{2}}$, intercepted by the co-ordinate axes. *Ans.* $\dfrac{\pi a^3}{15}$.

7. Find the volume and surface of the torus generated by revolving about X, the circle $x^2 + (y-b)^2 = a^2$.
Ans. $2\pi^2 a^2 b$ and $4\pi^2 ab$.

8. Find the volume and surface generated by revolving about Y, the catenary $y = \dfrac{a}{2}(e^{\frac{x}{a}} + e^{-\frac{x}{a}})$, from $x=0$ to $x=a$.
Ans. $\dfrac{\pi a^3}{2}(e + 5e^{-1} - 4)$ and $2\pi a^2(1 - e^{-1})$.

52. *Other Volumes.* The method of finding the volume of a solid of revolution in Art. 50, by considering it generated by a moving circle of varying radius, may be extended to any solid, where the area of a section can be expressed as a function of its perpendicular distance from a fixed point.

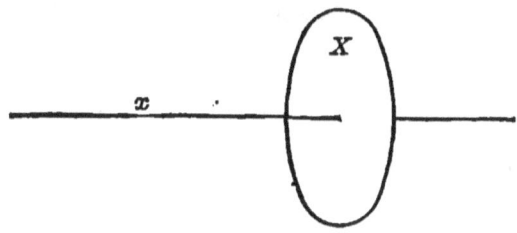

If we denote this distance by x, and the area of the section by X, we have for the volume,

$$V = \int X\, dx. \qquad \ldots \ldots \quad (1)$$

EXAMPLES.

1. Find the volume of a pyramid or cone having any base whatever.

   Let A be the area of the base, and h the altitude.

   Let x denote the perpendicular distance from the vertex, of a section parallel to the base. Calling the area of this section X, as in (1), we have by solid geometry,

   $$\frac{X}{A} = \frac{x^2}{h^2}, \qquad X = \frac{Ax^2}{h^2}.$$

   Substituting in (1),

   $$V = \frac{A}{h^2}\int_0^h x^2 dx = \frac{A}{h^2}\frac{h^3}{3} = \frac{Ah}{3}.$$

2. Find the volume of a right conoid with circular base, the radius of base being a, and altitude h.

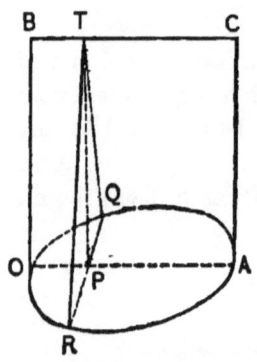

   $OA = BC = 2a, \quad BO = CA = h.$

   The section RTQ, perpendicular to OA, is an isosceles triangle.

   Let $x = OP$; then

   $X = \text{area } RTQ = PT \times PQ = h\sqrt{2ax - x^2}.$

   Substituting in (1), we have

   $$V = h\int_0^{2a} \sqrt{2ax - x^2}\, dx = \frac{\pi a^2 h}{2}.$$

   This is one-half the cylinder of the same base and altitude.

3. A rectangle moves from a fixed point, one side varying as the distance from this point, and the other as the square of this distance. At the distance of 2 feet, the rectangle becomes a square of 3 feet. What is the volume then generated? *Ans.* $4\tfrac{1}{2}$ cubic feet.

4. On the double ordinates of the ellipse $\frac{x^2}{a^2}+\frac{y^2}{b^2}=1$, and in planes perpendicular to that of the ellipse, isosceles triangles of vertical angle 2α are described. Find the volume of the surface thus constructed.

Ans. $\dfrac{4ab^2}{3\tan\alpha}$.

5. Given a right cylinder, altitude h, and radius of base a. Through a diameter of the upper base two planes are passed, touching the lower base on opposite sides. Find the volume included between the planes.

Ans. $\left(\pi-\dfrac{4}{3}\right)a^2h$.

6. Two cylinders of equal altitude h have a circle of radius a, for their common upper base. Their lower bases are tangent to each other. Find the volume common to the two cylinders.

Ans. $\dfrac{4a^2h}{3}$.

CHAPTER IX.

SUCCESSIVE INTEGRATION.

53. *Double Integral.* If we reverse the operations represented by $\dfrac{\partial^2 u}{\partial x\, \partial y}$, we have what is called a *double integral*.

For example, suppose $\dfrac{\partial^2 u}{\partial x\, \partial y} = x^2 y^3$,

then
$$u = \int\int x^2 y^3\, dy\, dx,$$

which indicates two successive integrations, the first with reference to x, regarding y as a constant, and the second with reference to y, regarding x as a constant. Thus

$$u = \int \frac{x^3 y^3}{3}\, dy = \frac{x^3 y^4}{12},$$

omitting the constants of integration.

54. *Definite Double Integral.* Here the integrations are between given limits.

For example,

$$\int_b^{2b}\int_0^a (a-x)y^2\, dy\, dx = \int_b^{2b}\left(ax - \frac{x^2}{2}\right)_0^a y^2\, dy$$

$$= \int_b^{2b} \frac{a^2}{2} y^2\, dy = \frac{7\, a^2 b^3}{6}.$$

In the above $\int_b^{2b}\int_0^a (a-x)y^2\, dy\, dx$, the right integral sign with the limits 0 and a, is to be used with the variable x, and the left with the limits b and $2b$, with the variable y; that is, the integral signs with their limits are to be taken in the same order as the differentials dy, dx, at the end, and from *right* to *left*.

55. Sometimes the limits of the first integration are functions of the variable of the second.

For example,

$$\int_0^a \int_{y-a}^{2y} xy\, dy\, dx = \int_0^a \left(\frac{x^2}{2}\right)_{y-a}^{2y} y\, dy = \frac{1}{2}\int_0^a (3y^3 + 2ay^2 - a^2y)\, dy$$
$$= \frac{11\, a^4}{24}.$$

As another example,

$$\int_0^a \int_0^{\sqrt{a^2-x^2}} (x+y)\, dx\, dy = \int_0^a \left(xy + \frac{y^2}{2}\right)_0^{\sqrt{a^2-x^2}} dx$$
$$= \int_0^a \left(x\sqrt{a^2-x^2} + \frac{a^2-x^2}{2}\right) dx = \frac{2\,a^3}{3}.$$

56. *Triple Integrals.* A similar notation is used for three successive integrations. Thus

$$\int_b^a \int_0^b \int_a^{2a} x^2 y^2 z\, dx\, dy\, dz = \int_b^a \int_0^b \frac{3a^2}{2} x^2 y^2\, dx\, dy$$
$$= \frac{3a^2}{2}\int_b^a \frac{b^3}{3} x^2\, dx = \frac{a^2 b^3}{2}\left(\frac{a^3}{3} - \frac{b^3}{3}\right) = \frac{a^2 b^3}{6}(a^3 - b^3).$$

EXAMPLES.

Evaluate the following definite integrals:

1. $\displaystyle\int_0^a \int_0^b xy(x-y)\, dx\, dy = \frac{a^2 b^2}{6}(a-b).$

2. $\displaystyle\int_b^a \int_\beta^a r^2 \sin\theta\, dr\, d\theta = \frac{a^3 - b^3}{3}(\cos\beta - \cos a).$

3. $\displaystyle\int_a^{2a} \int_y^{\frac{y^2}{a}} (x+y)\, dy\, dx = \frac{67\, a^3}{20}.$

4. $\displaystyle\int_{\frac{b}{2}}^{b} \int_0^{r} r\, dr\, d\theta = \frac{7\, b^2}{24}.$

5. $\displaystyle\int_0^\pi \int_0^{a(1+\cos\theta)} r^2 \sin\theta\, d\theta\, dr = \frac{4a^3}{3}.$

6. $\displaystyle\int_0^b \int_t^{10t} \sqrt{st-t^2}\, dt\, ds = 6b^3.$

7. $\displaystyle\int_a^{2a} \int_0^x \int_y^x xyz\, dx\, dy\, dz = \frac{21\,a^6}{16}.$

8. $\displaystyle\int_0^1 \int_0^x \int_0^{x+y} e^{x+y+z}\, dx\, dy\, dz = \frac{e^4-3}{8} - \frac{3e^2}{4} + e.$

CHAPTER X.

DOUBLE INTEGRATION APPLIED TO PLANE AREAS AND MOMENT OF INERTIA.

57. *Moment of Inertia.* As an illustration of double integration, we shall consider the problem of finding the moment of inertia of a given plane area.

Definition. The *moment of inertia* of a given plane area about a given point in the plane, is the sum of the products obtained, by multiplying the area of each infinitesimal portion by the square of its distance from the given point.

58. *Double Integration. Rectangular Co-ordinates.* To find the moment of inertia of the rectangle $OACB$ about O.

Let $OA = a$, $OB = b$.

Suppose the rectangle divided into rectangular elements by lines parallel to the co-ordinate axes. Let x, y, which are to be regarded as independent variables, be the co-ordinates of any point of intersection as P, and $x + dx$, $y + dy$, the co-ordinates of Q. Then the area of the element PQ is $dx\,dy$.

Moment of $PQ = \overline{OP}^2 \cdot dx\,dy = (x^2 + y^2)\,dx\,dy$.

The moment of the entire rectangle $OACB$ is the sum of all the terms obtained from $(x^2 + y^2)\,dx\,dy$, by varying x from 0 to a, and y from 0 to b.

If we suppose x to be constant, while y varies from 0 to b, we shall have the terms that constitute a vertical strip $MNN'M'$.

Hence

Moment of $MNN'M' = dx \int_0^b (x^2 + y^2) dy$

$= dx \left(x^2 y + \dfrac{y^3}{3} \right)_0^b = \left(bx^2 + \dfrac{b^3}{3} \right) dx.$

Having thus found the moment of a vertical strip, we may sum all these strips, by supposing x in this result to vary from 0 to a. That is,

Moment of $OACB = \int_0^a \left(bx^2 + \dfrac{b^3}{3} \right) dx = \dfrac{a^3 b + ab^3}{3}.$

But the preceding operations are the same as those represented by the double integral,

$$\int_0^a \int_0^b (x^2 + y^2) dx \, dy. \qquad \text{(See Art. 54.)}$$

If we first collect all the elements in a *horizontal* strip, and then sum these horizontal strips, we have

Moment of $OACB = \int_0^b \int_0^a (x^2 + y^2) dy \, dx = \dfrac{a^3 b + ab^3}{3}.$

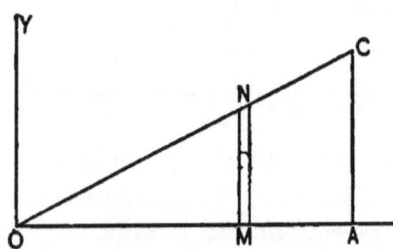

59. To find the moment of inertia of the right triangle OAC about O.

Let $OA = a$, $AC = b$. The equation of OC is

$$y = \dfrac{b}{a} x.$$

This differs from the preceding problem only in the limits of the first integration. In collecting the elements in a vertical strip MN, y varies from 0 to MN. But MN is no longer a constant as in Art. 58, but varies with OM, according to the equation of OC, $y = \dfrac{b}{a} x$. Hence the limits of y are 0 and $\dfrac{b}{a} x$.

In collecting all the vertical strips by the second integration, x varies from 0 to a, as in Art. 58.

APPLICATION OF DOUBLE INTEGRATION. 263

∴ Moment of $OAC = \int_0^a \int_0^{\frac{bx}{a}} (x^2 + y^2) dx\, dy = ab\left(\frac{a^2}{4} + \frac{b^2}{12}\right).$

By supposing the triangle composed of *horizontal* strips as HK, we shall find

Moment of OAC

$= \int_0^b \int_{\frac{ay}{b}}^{a} (x^2 + y^2) dy\, dx$

$= ab\left(\frac{a^2}{4} + \frac{b^2}{12}\right).$

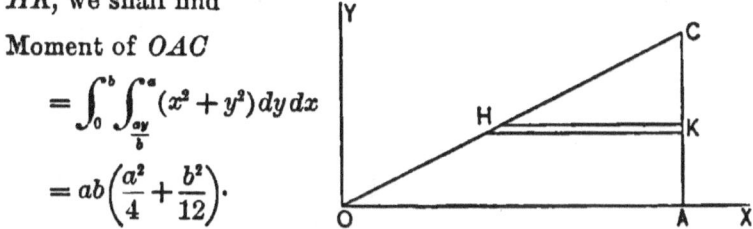

60. Plane Area as a Double Integral. If in Art. 58 we omit the factor $(x^2 + y^2)$, we shall have instead of the moment, the area, of the given surface.

That is, $\quad$ Area $= \iint dx\, dy = \iint dy\, dx,$

the limits being determined as before.

EXAMPLES.

1. Find the moment of inertia about the origin, of the right triangle formed by the co-ordinate axes and the line joining the points $(a, 0)$, $(0, b)$.

$$Ans. \quad \int_0^a \int_0^{\frac{b(a-x)}{a}} (x^2 + y^2) dx\, dy = \frac{ab(a^2 + b^2)}{12}.$$

2. Find the moment of inertia about the origin, of the circle $x^2 + y^2 = a^2$. $\quad Ans.\ 4 \int_0^a \int_0^{\sqrt{a^2 - x^2}} (x^2 + y^2) dx\, dy = \frac{\pi a^4}{2}.$

3. Find also the area of the preceding circle by Art. 60.

$$Ans.\ \pi a^2.$$

4. Find by Art. 60, the area between a straight line and a parabola, each of which joins the origin and the point (a, b), the axis of X being the axis of the parabola.

$$Ans. \quad \int_0^a \int_{\frac{bx}{a}}^{b\sqrt{\frac{x}{a}}} dx\, dy = \int_0^b \int_{\frac{ay^2}{b^2}}^{\frac{ay}{b}} dy\, dx = \frac{ab}{6}.$$

264 INTEGRAL CALCULUS.

5. Find the moment of inertia of the preceding area about the origin.

$$\text{Ans. } \frac{ab}{4}\left(\frac{a^2}{7} + \frac{b^2}{5}\right).$$

6. Find the moment of inertia about the origin, of the area included within the parabola $y^2 = 4ax$, the line $x + y = 3a$, and the axis of X.

$$\text{Ans. } \int_0^a \int_0^{2\sqrt{ax}} (x^2 + y^2)\,dx\,dy + \int_a^{3a} \int_0^{3a-x} (x^2 + y^2)\,dx\,dy$$

$$= \int_0^{2a} \int_{\frac{y^2}{4a}}^{3a-y} (x^2 + y^2)\,dy\,dx = \frac{314\,a^4}{35}.$$

61. *Double Integration. Polar Co-ordinates.* To find the area of the quadrant of a circle AOB, whose radius is a.

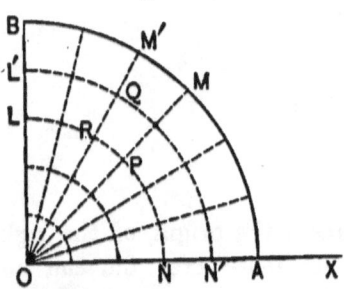

In rectangular co-ordinates, Art. 58, the lines of division consist of two systems, for one of which x is constant, and for the other, y is constant.

So in polar co-ordinates, we have one system of straight lines through the pole, for each of which θ is constant, and another system of circles about the pole as centre, for each of which r is constant.

Let r, θ, which are to be regarded as independent variables, be the co-ordinates of any point of intersection as P, and $r + dr$, $\theta + d\theta$, the co-ordinates of Q. Then the area of PQ is $PR \times RQ = r\,d\theta \cdot dr$.

If we first integrate regarding θ constant, while r varies from 0 to a, we collect all the elements in any sector MOM'.

The second integration sums all the sectors, by varying θ from 0 to $\frac{\pi}{2}$.

Hence Area $BOA = \displaystyle\int_0^{\frac{\pi}{2}} \int_0^a r\,d\theta\,dr = \frac{\pi a^2}{4}.$

APPLICATION OF DOUBLE INTEGRATION. 265

If we reverse the order of integration, integrating first with reference to θ, and afterwards with reference to r, we collect all the elements in a circular strip $NLL'N'$, and sum all these strips. This is written

$$\text{Area } BOA = \int_0^a \int_0^{\frac{\pi}{2}} r \, dr \, d\theta.$$

62. If the moment of inertia about O is required, we have for the moment of PQ, $r^2 \cdot r \, d\theta \, dr$. Hence,

$$\text{Moment of } BOA = \int_0^{\frac{\pi}{2}} \int_0^a r^3 \, d\theta \, dr = \int_0^a \int_0^{\frac{\pi}{2}} r^3 \, dr \, d\theta = \frac{\pi a^4}{8}.$$

63. To find by a double integration the area of the semicircle OBA with radius $OC = a$, the pole being on the circumference.

The polar equation of the circle is $r = 2a\cos\theta$. If we integrate first with reference to r, then with reference to θ, we shall have

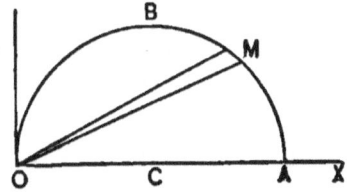

$$\text{Area } OBA = \int_0^{\frac{\pi}{2}} \int_0^{2a\cos\theta} r \, d\theta \, dr = \frac{\pi a^2}{2}.$$

Here, in collecting the elements in a radial strip OM, r varies from 0 to OM. But OM varies with θ, according to the equation of the circle $r = 2a\cos\theta$. Hence the limits are 0 and $2a\cos\theta$.

In collecting all these radial strips for the second integration, θ varies from 0 to $\frac{\pi}{2}$.

By supposing the area composed of concentric circular strips about O as LK, we find

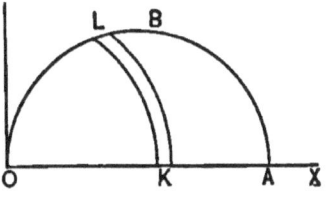

$$\text{Area } OBA = \int_0^{2a} \int_0^{\cos^{-1}\left(\frac{r}{2a}\right)} r \, dr \, d\theta = \frac{\pi a^2}{2}.$$

EXAMPLES.

1. Find the moment of inertia about O of the area of the semicircle in Art. 63.

    Ans. $\dfrac{3\pi a^4}{4}$.

2. Find the moment of inertia about the pole, of the area included by the parabola $r = a\sec^2\dfrac{\theta}{2}$, the initial line OX, and a line at right angles to it through the pole.

    Ans. $\displaystyle\int_0^{\frac{\pi}{2}}\int_0^{a\sec^2\frac{\theta}{2}} r^3\, d\theta\, dr = \dfrac{48\,a^4}{35}$.

3. Find the moment of inertia about its centre, of the area of one loop of the lemniscate $r^2 = a^2\cos 2\theta$.

    Ans. $\dfrac{\pi a^4}{16}$.

4. Find by double integration the entire area of the cardioid $r = a(1 - \cos\theta)$.

    Ans. $\dfrac{3\pi a^2}{2}$.

5. Find the moment of inertia about the pole, of the area of the preceding cardioid.

    Ans. $\dfrac{35\pi a^4}{16}$.

CHAPTER XI.

SURFACE AND VOLUME OF ANY SOLID.

64. *To find the area of any surface, whose equation is given between three rectangular co-ordinates, x, y, z.*

Let this equation be
$$z = f(x, y).$$

Suppose the given surface to be divided into elements by two series of planes, parallel respectively to XZ and YZ. These planes will also divide the plane XY into elementary

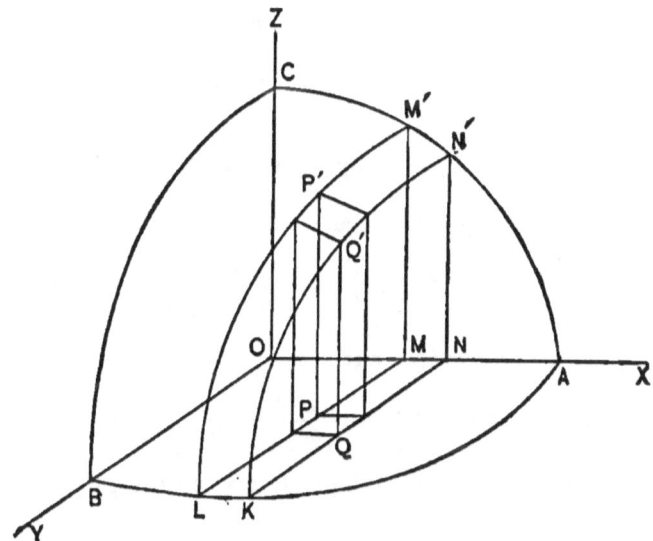

rectangles, one of which is PQ, the projection upon the plane XY of the corresponding element of the surface $P'Q'$.

Let x, y, z, be the co-ordinates of P', and $x + dx$, $y + dy$, $z + dz$, of Q'.

Since PQ is the projection of $P'Q'$, the area of PQ is equal to that of $P'Q'$, multiplied by the cosine of the inclination of $P'Q'$ to the plane XY. This angle is evidently that made by the tangent plane at P' with the plane XY. Denoting this angle by γ,

$$\text{Area } PQ = \text{Area } P'Q' \cdot \cos \gamma,$$
$$\text{Area } P'Q' = \text{Area } PQ \cdot \sec \gamma.$$

We see from the figure that

$$\text{Area } PQ = dx\, dy.$$

Also from analytical geometry of three dimensions,

$$\sec \gamma = \left[1 + \left(\frac{\partial z}{\partial x}\right)^2 + \left(\frac{\partial z}{\partial y}\right)^2\right]^{\frac{1}{2}}, \quad \text{(See p. 293.)}$$

where $\dfrac{\partial z}{\partial x}$ and $\dfrac{\partial z}{\partial y}$ are partial differential coefficients, taken from the equation of the given surface $z = f(x, y)$.

Hence $\quad\text{Area } P'Q' = \left[1 + \left(\dfrac{\partial z}{\partial x}\right)^2 + \left(\dfrac{\partial z}{\partial y}\right)^2\right]^{\frac{1}{2}} dx\, dy.$

If S denote the required surface,

$$S = \iint \left[1 + \left(\frac{\partial z}{\partial x}\right)^2 + \left(\frac{\partial z}{\partial y}\right)^2\right]^{\frac{1}{2}} dx\, dy, \quad . \quad . \quad (1)$$

the limits of the integration depending upon the projection, on the plane XY, of the surface required.

65. For example, suppose the surface ABC to be one-eighth of the surface of a sphere whose equation is

$$x^2 + y^2 + z^2 = a^2.$$

Here $\quad \dfrac{\partial z}{\partial x} = -\dfrac{x}{z}, \quad \dfrac{\partial z}{\partial y} = -\dfrac{y}{z}.$

$$1 + \left(\frac{\partial z}{\partial x}\right)^2 + \left(\frac{\partial z}{\partial y}\right)^2 = 1 + \frac{x^2}{z^2} + \frac{y^2}{z^2} = \frac{a^2}{z^2}.$$

Substituting in (1) Art. 64, we have

$$S = \iint \frac{a}{z} dx\, dy = a \iint \frac{dx\, dy}{\sqrt{a^2 - x^2 - y^2}}$$

Integrating first with reference to y, we collect all the elements in a strip $M'N'KL$, y varying from zero to ML, that is, between the limits 0 and $\sqrt{a^2 - x^2}$.

Integrating afterwards with reference to x, we sum all the strips, to obtain the required surface ABC, x varying from 0 to a.

Hence $\quad S = a \int_0^a \int_0^{\sqrt{a^2-x^2}} \frac{dx\, dy}{\sqrt{a^2 - x^2 - y^2}} = \frac{\pi a^2}{2}.$

EXAMPLES.

1. The axes of two equal right circular cylinders, a being the radius of base, intersect at right angles; find the surface of one intercepted by the other.

    Take for the equations of the cylinders,

    $$x^2 + z^2 = a^2, \text{ and } x^2 + y^2 = a^2.$$

    Ans. $8a \int_0^a \int_0^{\sqrt{a^2-x^2}} \frac{dx\, dy}{\sqrt{a^2 - x^2}} = 8a^2.$

2. The centre of a sphere, whose radius is a, is on the surface of a right circular cylinder, the radius of whose base is $\frac{a}{2}$. Find the surface of the sphere intercepted by the cylinder.

    Take for the equations of the sphere and cylinder,

    $$x^2 + y^2 + z^2 = a^2, \text{ and } x^2 + y^2 = ax.$$

    Ans. $4a \int_0^a \int_0^{\sqrt{ax-x^2}} \frac{dx\, dy}{\sqrt{a^2 - x^2 - y^2}} = 2(\pi - 2)a^2.$

3. In the preceding example, find the surface of the cylinder intercepted by the sphere.

    Ans. $2a \int_0^a \int_0^{\sqrt{a^2-ax}} \frac{dx\, dz}{\sqrt{ax - x^2}} = 4a^2.$

4. Find the area of that part of the surface

$$z^2 + (x\cos a + y\sin a)^2 = a^2,$$

which is situated in the positive compartment of co-ordinates.

The surface is a right circular cylinder, whose axis is the line $\quad z = 0, \quad x\cos a + y\sin a = 0,$
and radius of base a. $\qquad$ Ans. $\dfrac{a^2}{\sin a \cos a}$.

5. A diameter of a sphere, whose radius is a, is the axis of a right prism with a square base, $2b$ being the side of the square. Find the surface of the sphere intercepted by the prism. $\quad$ Ans. $8a\left(2b\sin^{-1}\dfrac{b}{\sqrt{a^2-b^2}} - a\sin^{-1}\dfrac{b^2}{a^2-b^2}\right)$.

66. *To find the volume of any solid bounded by a surface, whose equation is given between three rectangular co-ordinates, x, y, z.*

As a plane area, by dividing it into elementary rectangles, is

$$A = \iint dx\, dy,$$

so any solid may be supposed to be divided, by planes parallel to the co-ordinate planes, into elementary rectangular parallelopipeds. The volume of one of these parallelopipeds is $dx\, dy\, dz$, and the volume of the entire solid is

$$V = \iiint dx\, dy\, dz,$$

the limits of the integration depending upon the equation of the bounding surface.

VOLUME OF ANY SOLID.

67. For example, let us find the volume of one-eighth of the ellipsoid, whose equation is

$$\frac{x^2}{a^2}+\frac{y^2}{b^2}+\frac{z^2}{c^2}=1.$$

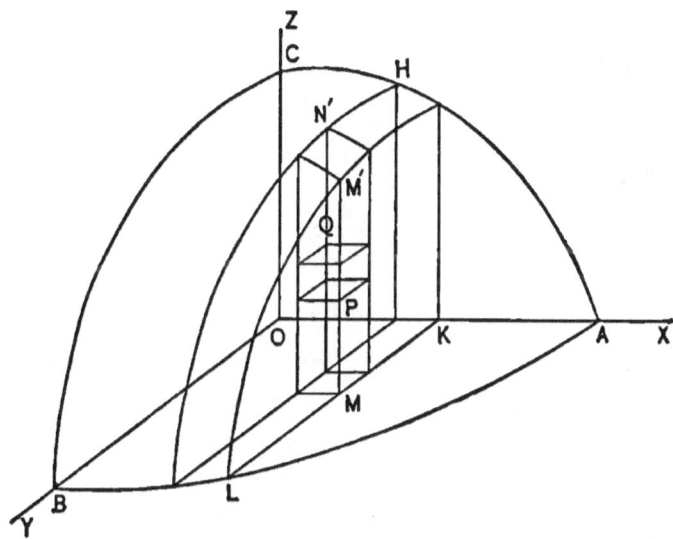

PQ represents one of the elementary parallelopipeds whose volume is $dx\,dy\,dz$.

If we integrate with reference to z, we collect all the elements in the column MN', z varying from zero to MM'; that is, from 0 to $z = c\sqrt{1-\frac{x^2}{a^2}-\frac{y^2}{b^2}}$.

Integrating next with reference to y, we collect all the columns in the slice $KLN'H$, y varying from zero to KL; that is, from 0 to $y = b\sqrt{1-\frac{x^2}{a^2}}$.

This value of y is taken from the equation of the curve ALB,

$$\frac{x^2}{a^2}+\frac{y^2}{b^2}=1.$$

Finally, we integrate with reference to x, to collect all the slices in the entire solid ABC. Here x varies from zero to OA; that is, from 0 to a.

Hence we have

$$V = \int_0^a \int_0^{b\sqrt{1-\frac{x^2}{a^2}}} \int_0^{c\sqrt{1-\frac{x^2}{a^2}-\frac{y^2}{b^2}}} dx\,dy\,dz.$$

Evaluating this integral, we find

$$V = \frac{\pi abc}{6}.$$

For the entire ellipsoid,

$$V = \frac{4\pi abc}{3}.$$

EXAMPLES.

1. Find the volume of one of the wedges cut from the cylinder, $x^2 + y^2 = a^2$, by the planes

$$z = 0 \quad \text{and} \quad z = x\tan\alpha.$$

Ans. $2\int_0^a \int_0^{\sqrt{a^2-x^2}} \int_0^{x\tan\alpha} dx\,dy\,dz = \dfrac{2a^3 \tan\alpha}{3}$.

2. Find the volume of the solid contained between the paraboloid of revolution

$$x^2 + y^2 = az,$$

the cylinder $\quad x^2 + y^2 = 2ax,$

and the plane $\quad z = 0.$

Ans. $2\int_0^{2a} \int_0^{\sqrt{2ax-x^2}} \int_0^{\frac{x^2+y^2}{a}} dx\,dy\,dz = \dfrac{3\pi a^3}{2}$.

3. Find the volume bounded by the surface

$$\left(\frac{x}{a}\right)^{\frac{1}{2}} + \left(\frac{y}{b}\right)^{\frac{1}{2}} + \left(\frac{z}{c}\right)^{\frac{1}{2}} = 1,$$

and by the positive sides of the three co-ordinate planes.

Ans. $\dfrac{abc}{90}$.

4. The centre of a sphere of radius a, is on the surface of a right circular cylinder, the radius of whose base is $\frac{a}{2}$. Find the volume of the part of the cylinder intercepted by the sphere. (See Ex. 2, Art. 65.)

$$\text{Ans. } \frac{2}{3}\left(\pi - \frac{4}{3}\right)a^3.$$

5. Find the entire volume bounded by the surface, whose equation is $x^{\frac{2}{3}} + y^{\frac{2}{3}} + z^{\frac{2}{3}} = a^{\frac{2}{3}}$. $\quad$ Ans. $\dfrac{4\pi a^3}{35}$.

CHAPTER XII.

HYPERBOLIC FUNCTIONS. EQUATIONS AND PROPERTIES OF CYCLOID, EPICYCLOID, AND HYPOCYCLOID. INTRINSIC EQUATION OF A CURVE.

68. We have reserved for this chapter certain miscellaneous subjects, for the treatment of which, both the Differential, and Integral, Calculus are required.

HYPERBOLIC FUNCTIONS.

69. *Definitions.* By analogy with the exponential values of the sine and cosine, on page 60,

$$\sin x = \frac{e^{x\sqrt{-1}} - e^{-x\sqrt{-1}}}{2\sqrt{-1}}, \qquad \cos x = \frac{e^{x\sqrt{-1}} + e^{-x\sqrt{-1}}}{2}; \quad . \quad . \quad (1)$$

the real functions

$$\frac{e^x - e^{-x}}{2}, \quad \text{and} \quad \frac{e^x + e^{-x}}{2},$$

are called the *hyperbolic sine*, and *hyperbolic cosine*, of x, and written

$$\sinh x = \frac{e^x - e^{-x}}{2}, \qquad \cosh x = \frac{e^x + e^{-x}}{2}.$$

By substituting $x\sqrt{-1}$ for x in (1), we find

$$\sinh x = \frac{\sin(x\sqrt{-1})}{\sqrt{-1}}, \qquad \cosh x = \cos(x\sqrt{-1}).$$

It is evident also that

$$\sinh 0 = 0, \qquad \cosh 0 = 1,$$
$$\sinh(-x) = -\sinh x, \qquad \cosh(-x) = \cosh x.$$

HYPERBOLIC FUNCTIONS. 275

The functions, $\sinh x$, $\cosh x$, for real values of x, are not periodic functions like $\sin x$, $\cos x$, but increase with x to infinity.

The other hyperbolic functions are

$$\tanh x = \frac{\sinh x}{\cosh x} = \frac{e^x - e^{-x}}{e^x + e^{-x}},$$

$$\coth x = \frac{1}{\tanh x} = \frac{e^x + e^{-x}}{e^x - e^{-x}},$$

$$\operatorname{sech} x = \frac{1}{\cosh x} = \frac{2}{e^x + e^{-x}},$$

$$\operatorname{cosech} x = \frac{1}{\sinh x} = \frac{2}{e^x - e^{-x}}.$$

70. From these definitions we find

$$\cosh^2 x - \sinh^2 x = 1,$$
$$\tanh^2 x + \operatorname{sech}^2 x = 1,$$
$$\coth^2 x - \operatorname{cosech}^2 x = 1,$$
$$\sinh 2x = 2 \sinh x \cosh x,$$
$$\cosh 2x = \cosh^2 x + \sinh^2 x,$$
$$\sinh(x \pm y) = \sinh x \cosh y \pm \cosh x \sinh y,$$
$$\cosh(x \pm y) = \cosh x \cosh y \pm \sinh x \sinh y,$$
$$\tanh(x \pm y) = \frac{\tanh x \pm \tanh y}{1 \pm \tanh x \tanh y}.$$

71. *Inverse Hyperbolic Functions.*

If $\qquad x = \sinh y,$ . . . . (1)

then $\qquad y = \sinh^{-1} x.$

But from (1),

$$x = \frac{e^y - e^{-y}}{2}.$$

Solving this with reference to y,
$$y = \log(x + \sqrt{x^2+1}).$$

Hence $\quad \sinh^{-1} x = \log(x + \sqrt{x^2+1}).$

Similarly, $\cosh^{-1} x = \log(x + \sqrt{x^2-1}).$

$$\tanh^{-1} x = \frac{1}{2}\log\frac{1+x}{1-x},$$

$$\coth^{-1} x = \tanh^{-1}\frac{1}{x} = \frac{1}{2}\log\frac{x+1}{x-1},$$

$$\operatorname{sech}^{-1} x = \cosh^{-1}\frac{1}{x} = \log\frac{1+\sqrt{1-x^2}}{x},$$

$$\operatorname{cosech}^{-1} x = \sinh^{-1}\frac{1}{x} = \log\frac{1+\sqrt{1+x^2}}{x}.$$

72. *Differentiation of Hyperbolic Functions.* From the definitions we have

$$\frac{d}{dx}\sinh x = \cosh x,$$

$$\frac{d}{dx}\cosh x = \sinh x,$$

$$\frac{d}{dx}\tanh x = \operatorname{sech}^2 x,$$

$$\frac{d}{dx}\coth x = -\operatorname{cosech}^2 x,$$

$$\frac{d}{dx}\operatorname{sech} x = -\operatorname{sech} x \tanh x,$$

$$\frac{d}{dx}\operatorname{cosech} x = -\operatorname{cosech} x \coth x.$$

To differentiate the inverse function
$$y = \sinh^{-1} x,$$
we have $\quad x = \sinh y,$
$$\frac{dx}{dy} = \cosh y = \sqrt{\sinh^2 y + 1} = \sqrt{x^2+1},$$

$$\frac{dy}{dx} = \frac{1}{\sqrt{x^2+1}}.$$

Hence $\quad \dfrac{d}{dx}\sinh^{-1}x = \dfrac{1}{\sqrt{x^2+1}}.$

Similarly, $\quad \dfrac{d}{dx}\cosh^{-1}x = \dfrac{1}{\sqrt{x^2-1}},$

$$-\frac{d}{dx}\tanh^{-1}x = \frac{1}{1-x^2},$$

$$\frac{d}{dx}\coth^{-1}x = \frac{1}{1-x^2},$$

$$-\frac{d}{dx}\operatorname{sech}^{-1}x = -\frac{1}{x\sqrt{1-x^2}},$$

$$-\frac{d}{dx}\operatorname{cosech}^{-1}x = -\frac{1}{x\sqrt{1+x^2}}.$$

73. Inverse Circular, and Inverse Hyperbolic, Functions as Integrals. A comparison of the integrals involving the inverse circular functions, with those involving the inverse hyperbolic functions, shows the close analogy between them.

$$\int \frac{dx}{\sqrt{a^2-x^2}} = \sin^{-1}\frac{x}{a}, \qquad \int \frac{dx}{\sqrt{x^2+a^2}} = \sinh^{-1}\frac{x}{a}.$$

$$\text{or } = -\cos^{-1}\frac{x}{a}. \qquad \int \frac{dx}{\sqrt{x^2-a^2}} = \cosh^{-1}\frac{x}{a}.$$

$$\int \frac{dx}{a^2+x^2} = \frac{1}{a}\tan^{-1}\frac{x}{a}, \qquad \int \frac{dx}{a^2-x^2} = \frac{1}{a}\tanh^{-1}\frac{x}{a},$$

$$\text{or } = -\frac{1}{a}\cot^{-1}\frac{x}{a}. \qquad \text{or } = \frac{1}{a}\coth^{-1}\frac{x}{a}.$$

$$\int \frac{dx}{x\sqrt{x^2-a^2}} = \frac{1}{a}\sec^{-1}\frac{x}{a}, \qquad \int \frac{dx}{x\sqrt{a^2-x^2}} = -\frac{1}{a}\operatorname{sech}^{-1}\frac{x}{a}.$$

$$\text{or } = -\frac{1}{a}\operatorname{cosec}^{-1}\frac{x}{a}. \qquad \int \frac{dx}{x\sqrt{a^2+x^2}} = -\frac{1}{a}\operatorname{cosech}^{-1}\frac{x}{a}.$$

74. Circular and Hyperbolic Functions, as related to the Circle and Equilateral Hyperbola. To show the origin of the term *hyperbolic functions*, let us first consider the circle

$$x^2 + y^2 = a^2.$$

If we let

$$\theta = \text{angle } POA,$$

and $u =$ sectorial area POA,

we have

$$x = a\cos\theta, \quad y = a\sin\theta, \quad u = \frac{a^2\theta}{2}.$$

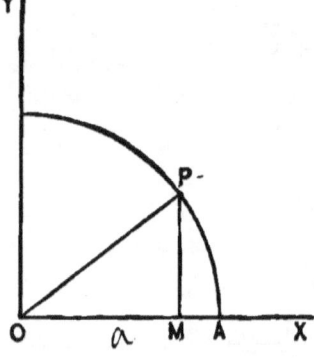

Hence
$$\left. \begin{array}{l} OM = x = a\cos\dfrac{2u}{a^2}, \\[4pt] PM = y = a\sin\dfrac{2u}{a^2}. \end{array} \right\} \quad \cdots \cdots (1)$$
and

We shall now show that if "cos" and "sin" in (1) are replaced by "cosh" and "sinh," then (1) will apply to the equilateral hyperbola

$$x^2 - y^2 = a^2. \quad \cdots (2)$$

Here the sectorial area POA is

$$u = \frac{a^2}{2}\log\frac{x+y}{a}.$$

(See Ex. 3, p. 246.)

Whence
$$\frac{x+y}{a} = e^{\frac{2u}{a^2}}. \quad \cdots \cdots (3)$$

From (2) and (3), $\quad \dfrac{x-y}{a} = e^{-\frac{2u}{a^2}}.$

Hence $\dfrac{x}{a} = \dfrac{e^{\frac{2u}{a^2}} + e^{-\frac{2u}{a^2}}}{2} = \cosh\dfrac{2u}{a^2}$,

and $\dfrac{y}{a} = \dfrac{e^{\frac{2u}{a^2}} - e^{-\frac{2u}{a^2}}}{2} = \sinh\dfrac{2u}{a^2}$.

Hence $OM = x = a\cosh\dfrac{2u}{a^2}$,

and $PM = y = a\sinh\dfrac{2u}{a^2}$,

which are similar expressions to (1).

If θ = angle POA, in the hyperbola,

$$\tan\theta = \dfrac{y}{x} = \tanh\dfrac{2u}{a^2};$$

hence $u = \dfrac{a^2}{2}\tanh^{-1}\tan\theta;$

whereas in the circle,

$$u = \dfrac{a^2}{2}\theta = \dfrac{a^2}{2}\tan^{-1}\tan\theta.$$

75. *Exercises in Hyperbolic Functions.*

1. $\tanh^{-1}\dfrac{2x}{1+x^2} = 2\tanh^{-1}x$.
2. $\sinh^{-1}(3x + 4x^3) = 3\sinh^{-1}x$.
3. $\tanh^{-1}\sin x = \operatorname{sech}^{-1}\cos x$.
4. $\tan^{-1}\sinh x = \sec^{-1}\cosh x$.
5. $2\tan^{-1}\tanh x = \tan^{-1}\sinh 2x$.
6. $2\tanh^{-1}\tan x = \tanh^{-1}\sin 2x$.
7. $2\cosh^{-1}\cos x = \cosh^{-1}\cos 2x$.
8. $2\cos^{-1}\cosh x = \cos^{-1}\cosh 2x$.

9. $y = \tan^{-1} x + \tanh^{-1} x.$ $\quad \dfrac{dy}{dx} = \dfrac{2}{1-x^4}.$

10. $y = \tan^{-1} \tanh x.$ $\quad \dfrac{dy}{dx} = \operatorname{sech} 2x.$

11. $y = \sinh^{-1} \tan x.$ $\quad \dfrac{dy}{dx} = \sec x.$

12. $y = \sin^{-1}\sqrt{\sin 2x} - \sinh^{-1}\sqrt{\sin 2x}.$ $\quad \dfrac{dy}{dx} = \sqrt{2 \cot x}.$

13. $y = \tan^{-1}\sqrt{\tanh x} + \tanh^{-1}\sqrt{\tanh x}.$ $\quad \dfrac{dy}{dx} = \sqrt{\coth x}.$

14. $\sinh x = x + \dfrac{x^3}{\lfloor 3} + \dfrac{x^5}{\lfloor 5} + \cdots.$

15. $\cosh x = 1 + \dfrac{x^2}{\lfloor 2} + \dfrac{x^4}{\lfloor 4} + \cdots.$

16. $\tanh^{-1} x = x + \dfrac{x^3}{3} + \dfrac{x^5}{5} + \cdots.$

17. Express the equation of the catenary $y = \dfrac{a}{2}(e^{\frac{x}{a}} + e^{-\frac{x}{a}})$, and also the length of the arc from the vertex, in hyperbolic functions.

Ans. $y = a \cosh \dfrac{x}{a}$, and $s = a \sinh \dfrac{x}{a}.$

EQUATION AND PROPERTIES OF THE CYCLOID.

76. *Definition*. The cycloid is the curve described by a point in the circumference of a circle, as it rolls along a straight line.

Let OX be the straight line. As the circle NPT, with radius a, rolls along this line, the point P describes the cycloid OBO'.

CYCLOID.

Let the angle through which the circle has rolled from O,
$$PCN = \theta;$$
and let x, y, be the co-ordinates of P.

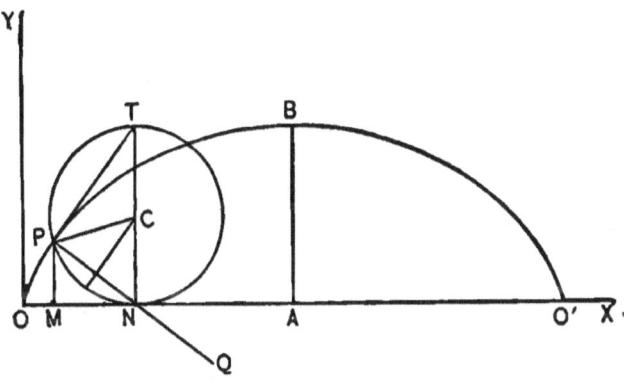

As P is supposed to have been in contact at O, it follows that
$$ON = \text{arc } PN = a\theta.$$
Then
$$\left. \begin{array}{l} x = OM = ON - MN = a\theta - a\sin\theta, \\ y = PM = CN - a\cos\theta = a - a\cos\theta. \end{array} \right\} \quad \ldots \quad (1)$$

If we eliminate θ between these equations, we have
$$x = a\cos^{-1}\frac{a-y}{a} - \sqrt{2ay - y^2},$$
or
$$x = a\text{ vers}^{-1}\frac{y}{a} - \sqrt{2ay - y^2}. \quad \ldots \ldots (2)$$

This is the equation of the cycloid, but equations (1) are generally more useful than (2).

77. The point B is called the vertex of the curve. If the origin is transferred from O to B with parallel axes, we have, x', y', being the new co-ordinates,
$$y = y' + 2a, \qquad x = x' + \pi a.$$

282 INTEGRAL CALCULUS.

Substituting these in (1), we obtain

$$x' = a(\theta - \pi) - a\sin\theta,$$
$$y' = -a - a\cos\theta.$$

Letting $\theta - \pi = \theta'$, the angle through which the circle has rolled from A, and omitting the accents on x' and y', we have

$$\left.\begin{array}{l} x = a\theta' + a\sin\theta', \\ y = -a + a\cos\theta', \end{array}\right\} \quad \ldots \ldots \ldots \quad (1)$$

the equation of the cycloid referred to its vertex.

78. Tangent and Normal. From (1) Art. 76, we have

$$\frac{dx}{d\theta} = a(1 - \cos\theta) = 2a\sin^2\frac{\theta}{2}, \quad \ldots \ldots \quad (1)$$

$$\frac{dy}{d\theta} = a\sin\theta = 2a\sin\frac{\theta}{2}\cos\frac{\theta}{2};$$

therefore $\quad \dfrac{dy}{dx} = \tan\phi = \cot\dfrac{\theta}{2}. \quad \ldots \ldots \ldots \quad (2)$

Hence $\quad \phi = \dfrac{\pi}{2} - \dfrac{\theta}{2}.$

But since $PTN = \dfrac{\theta}{2}$, the angle made by PT with the axis of X is $\dfrac{\pi}{2} - \dfrac{\theta}{2}$; hence PT is the tangent to the curve, and PN the normal.

79. Radius of Curvature. From (1) and (2) of the preceding article, we find

$$\frac{d^2y}{dx^2} = -\csc^2\frac{\theta}{2} \cdot \frac{1}{2}\frac{d\theta}{dx} = -\frac{\csc^2\dfrac{\theta}{2}}{4a\sin^2\dfrac{\theta}{2}} = -\frac{1}{4a\sin^4\dfrac{\theta}{2}}.$$

CYCLOID. 283

Substituting in the expression for the radius of curvature, we have

$$\rho = -\left(1+\cot^2\frac{\theta}{2}\right)^{\frac{3}{2}} 4a\sin^4\frac{\theta}{2} = -4a\sin\frac{\theta}{2} = -2PN.$$

Hence if we produce PN to Q, making $NQ = PN$, Q will be the centre of curvature for the point P.

80. Evolute. Produce the diameter TN, making $NR = TN$, and on NR as diameter describe the circle NR. This circle will pass through Q, since $NQ = PN$.

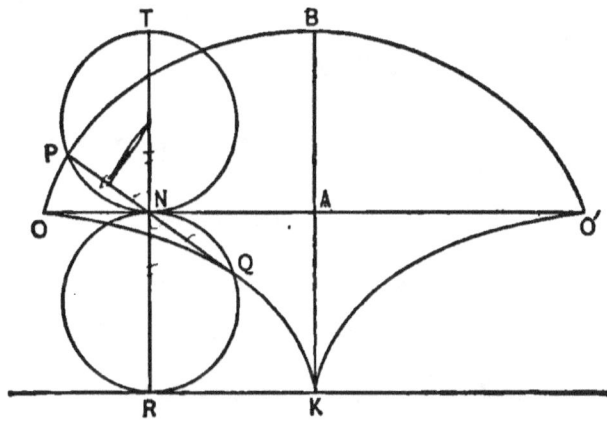

The $\qquad$ arc NQ = arc PN = ON,

and $\qquad$ arc $NQR = OA$;

therefore $\qquad$ arc $QR = OA - ON = RK$.

Hence Q is a point in an equal cycloid, generated by rolling the circle NQR from K along the straight line KR.

Hence the evolute of the cycloid OBO' is composed of the two semi-cycloids OK and KO'.

81. Length of Arc. To find the length of the arc OP (Fig. of Art. 76) we substitute in

$$s = \int \left[1 + \left(\frac{dy}{dx}\right)^2\right]^{\frac{1}{2}} dx,$$

$$\frac{dy}{dx} = \cot\frac{\theta}{2}, \quad \text{and} \quad dx = 2a\sin^2\frac{\theta}{2}d\theta.$$

We thus obtain

$$s = 2a\int_0^\theta \sin\frac{\theta}{2}d\theta = 4a\left(1 - \cos\frac{\theta}{2}\right).$$

If $\theta = 2\pi$, we have for the entire arc, $OBO' = 8a$.

This result is also evident from the property of the evolute, from which

$$OQK = BK = 4a.$$

82. Area. To find the area between the curve and the axis of X, we substitute in

$$A = \int y\,dx,$$

$$y = a(1 - \cos\theta), \quad dx = a(1 - \cos\theta)d\theta.$$

Thus we have for the entire area $OBO'A$,

$$A = a^2 \int_0^{2\pi} (1 - \cos\theta)^2 d\theta = 3\pi a^2.$$

Hence this area is three times that of the generating circle.

EPICYCLOID AND HYPOCYCLOID.

83. Equation of Epicycloid. The epicycloid is the curve described by a point in the circumference of a circle, which rolls outside of a fixed circle.

Suppose the circle BPS rolls on the fixed circle ADA', the point P describing the epicycloid APA'.

Let $OB = a$, $BC = b$, $BOA = \phi$, $BCP = \psi$.

Since the arcs BA and BP are equal, we have

$$a\phi = b\psi.$$

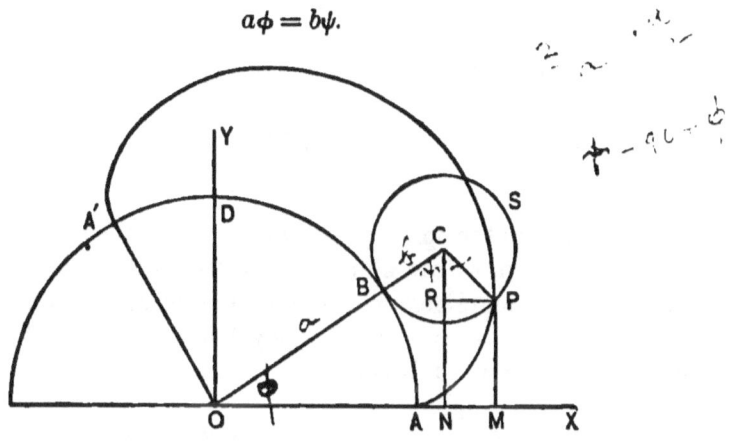

$$\begin{aligned}
x = OM &= ON + RP \\
&= (a+b)\cos\phi + b\sin\left[\psi - \left(\frac{\pi}{2} - \phi\right)\right] \\
&= (a+b)\cos\phi - b\cos(\psi + \phi). \quad \ldots \quad (1)
\end{aligned}$$

$$\begin{aligned}
y = PM &= CN - CR \\
&= (a+b)\sin\phi - b\cos\left[\psi - \left(\frac{\pi}{2} - \phi\right)\right] \\
&= (a+b)\sin\phi - b\sin(\psi + \phi). \quad \ldots \quad (2)
\end{aligned}$$

Substituting in (1) and (2), $\psi = \dfrac{a\phi}{b}$,

$$\left.\begin{aligned}
x &= (a+b)\cos\phi - b\cos\frac{a+b}{b}\phi, \\
y &= (a+b)\sin\phi - b\sin\frac{a+b}{b}\phi.
\end{aligned}\right\} \quad \ldots \quad (3)$$

84. Equation of Hypocycloid. The hypocycloid is the curve described by a point in the circumference of a circle, which rolls *inside* of a fixed circle.

If in equations (3) Art. 83, we change b into $-b$, we have the equations of the hypocycloid,

$$\left.\begin{array}{l} x = (a-b)\cos\phi + b\cos\dfrac{a-b}{b}\phi, \\ y = (a-b)\sin\phi - b\sin\dfrac{a-b}{b}\phi. \end{array}\right\} \quad \ldots \ldots \quad (1)$$

85. When, in the epicycloid or hypocycloid, the ratio between a and b is given, we can eliminate ϕ between the two equations, and obtain a single algebraic equation between x and y.

For example, consider the hypocycloid where $a = 4b$. Then equations (1) Art. 84, become

$$x = \frac{3a}{4}\cos\phi + \frac{a}{4}\cos 3\phi = a\cos^3\phi,$$

$$y = \frac{3a}{4}\sin\phi - \frac{a}{4}\sin 3\phi = a\sin^3\phi.$$

Whence $\quad x^{\frac{2}{3}} + y^{\frac{2}{3}} = a^{\frac{2}{3}}, \quad$ as given on page 96.

86. *Radius of Curvature of Epicycloid.* By differentiating (3) Art. 83, we have

$$\frac{dx}{d\phi} = (a+b)\left(\sin\frac{a+b}{b}\phi - \sin\phi\right) \quad \ldots \ldots \quad (1)$$

$$= 2(a+b)\sin\frac{a}{2b}\phi\cos\frac{a+2b}{2b}\phi. \quad \ldots \ldots \quad (2)$$

$$\frac{dy}{d\phi} = (a+b)\left(-\cos\frac{a+b}{b}\phi + \cos\phi\right) \quad \ldots \ldots \quad (3)$$

$$= 2(a+b)\sin\frac{a}{2b}\phi\sin\frac{a+2b}{2b}\phi. \quad \ldots \ldots \quad (4)$$

Therefore

$$\frac{dy}{dx} = \tan\frac{a+2b}{2b}\phi.$$

Whence

$$\frac{d^2y}{dx^2} = \frac{a+2b}{2b}\sec^2\frac{a+2b}{2b}\phi \cdot \frac{d\phi}{dx} = \frac{a+2b}{4b(a+b)} \frac{\sec^3\frac{a+2b}{2b}\phi}{\sin\frac{a}{2b}\phi}.$$

Substituting in the formula for the radius of curvature, we find

$$\rho = \frac{\left(1+\tan^2\frac{a+2b}{2b}\phi\right)^{\frac{3}{2}}}{\sec^3\frac{a+2b}{2b}\phi} \cdot \frac{4b(a+b)}{a+2b}\sin\frac{a}{2b}\phi$$

$$= \frac{4b(a+b)}{a+2b}\sin\frac{a}{2b}\phi = \frac{4b(a+b)}{a+2b}\sin\frac{\psi}{2}. \quad \dots \quad (5)$$

If $a = \infty$, the epicycloid becomes the cycloid, and

$$\frac{a+b}{a+2b} = 1.$$

Hence $\quad \rho = 4b\sin\frac{\psi}{2}$, as in Art. 79.

87. *Radius of Curvature of Hypocycloid.* By changing b into $-b$ in (5) Art. 86, we have for the radius of curvature of the hypocycloid, numerically,

$$\rho = \frac{4b(a-b)}{a-2b}\sin\frac{\psi}{2}.$$

88. *Length of Arc.* From (2) and (4), Art. 86, we have

$$\left(\frac{ds}{d\phi}\right)^2 = \left(\frac{dx}{d\phi}\right)^2 + \left(\frac{dy}{d\phi}\right)^2 = 4(a+b)^2\sin^2\frac{a}{2b}\phi.$$

Hence for the entire loop APA' (Fig. Art. 83), we have

$$s = 2(a+b)\int_0^{\frac{2\pi b}{a}}\sin\frac{a}{2b}\phi\, d\phi = \frac{8(a+b)b}{a}.$$

For the hypocycloid, the length of one loop is

$$s = \frac{8(a-b)b}{a}.$$

89. Area betweeen Curve and Fixed Circle. To find the area $APA'BA$ (Fig. Art. 83), it is better to use polar co-ordinates, r, θ. The formula,

$$A = \frac{1}{2}\int r^2 d\theta,$$

will give the area $APA'OA$, and this, less the area of the sector $A'OA$, will be the required area.

Differentiating $\dfrac{y}{x} = \tan\theta$,

we have $\dfrac{x\,dy - y\,dx}{x^2} = \sec^2\theta\,d\theta,$

$$x\,dy - y\,dx = x^2 \sec^2\theta\,d\theta = r^2 d\theta.$$

From (3) Art. 83, and (1), (3), Art. 86, we find

$$x\,dy - y\,dx = (a+b)(a+2b)\left(1 - \cos\frac{a}{b}\phi\right)d\phi.$$

Therefore

$$\int r^2 d\theta = (a+b)(a+2b)\int\left(1 - \cos\frac{a}{b}\phi\right)d\phi.$$

Hence

$$\text{Area } APA'OA = \frac{1}{2}(a+b)(a+2b)\int_0^{\frac{2\pi b}{a}}\left(1 - \cos\frac{a}{b}\phi\right)d\phi$$

$$= \frac{\pi b}{a}(a+b)(a+2b).$$

Subtracting the area of the sector

$$AOA' = \pi ab,$$

we have

$$\text{Area } APA'BA = \pi b\left[\frac{(a+b)(a+2b) - a^2}{a}\right] = \frac{\pi b^2(3a+2b)}{a}.$$

The corresponding area for the hypocycloid is

$$\frac{\pi b^2(3a-2b)}{a}.$$

INTRINSIC EQUATION OF A CURVE.

90. Definition. In considering the subject of curvature in Art. 111, page 120, the linear motion of a point along a curve is compared with the corresponding change of direction.

An equation expressing the relation between these quantities is called the *intrinsic equation* of the curve. It may be more precisely defined as follows:

The intrinsic equation of a curve is the relation between the length of the arc measured from some fixed point, and the angle by which its tangent deviates from the original direction at the fixed point.

It is called the intrinsic equation, because it is independent of any co-ordinate axes, or any external points or lines of reference.

Suppose O to be the point of the curve from which the arc is measured, and let OT be the tangent at O. Taking P as any point of the curve, and letting

$$s = \text{arc } OP,$$
and $\quad \phi = PMT,$

the intrinsic equation will be of the form

$$s = f(\phi).$$

The intrinsic equation of the circle whose radius is a is evidently $\quad s = a\phi.$

91. To find the intrinsic equation of a curve whose equation is given in rectangular or polar co-ordinates, it is only necessary to find the general expressions for s and ϕ, and eliminate the other variables.

For example, let us find the intrinsic equation of the cycloid.

Taking the vertex as origin, we use equations (1) Art. 77, reversing the direction of the axis of Y. We then have, omitting accents,
$$x = a(\theta + \sin\theta),$$
$$y = a(1 - \cos\theta).$$

Differentiating these equations, we obtain
$$\tan\phi = \frac{dy}{dx} = \frac{\sin\theta}{1+\cos\theta} = \tan\frac{\theta}{2}.$$

Hence $\quad\quad\phi = \frac{\theta}{2}.\quad\cdot\quad\cdot\quad\cdot\quad\cdot\quad\cdot\quad\cdot\quad\cdot\quad\cdot\quad$ (1)

Also $\quad\left(\dfrac{ds}{d\theta}\right)^2 = \left(\dfrac{dx}{d\theta}\right)^2 + \left(\dfrac{dy}{d\theta}\right)^2 = a^2(2 + 2\cos\theta) = 4a^2\cos^2\dfrac{\theta}{2}.$

Hence $\quad\quad s = 2a\displaystyle\int_0^\theta \cos\frac{\theta}{2}d\theta = 4a\sin\frac{\theta}{2}.\quad\cdot\quad\cdot\quad\cdot\quad$ (2)

Eliminating θ between (1) and (2), we have
$$s = 4a\sin\phi,$$
which is the intrinsic equation of the cycloid, referred to its vertex.

92. *Intrinsic Equation of the Evolute.*

If we differentiate the intrinsic equation of the curve
$$s = f(\phi),$$
we have, by (1) Art. 114, Dif. Cal., the radius of curvature,
$$\rho = \frac{ds}{d\phi} = f'(\phi). \quad\cdot\quad\cdot\quad\cdot\quad\cdot\quad\cdot\quad\cdot\quad\text{(1)}$$

Let O', P', be the centres of curvature for O, P, respectively, and $O'P'$, the evolute of OP.

Let $\quad\quad\quad s = OP, \quad\quad \phi = PMT,$
and $\quad\quad\quad s' = O'P', \quad\quad \phi' = P'M'T'.$

INTRINSIC EQUATION OF THE EVOLUTE.

Since tangents to $O'P'$ are normals to OP,
$$\phi' = \phi.$$
Also $\quad s' = O'P' = PP' - OO'.$

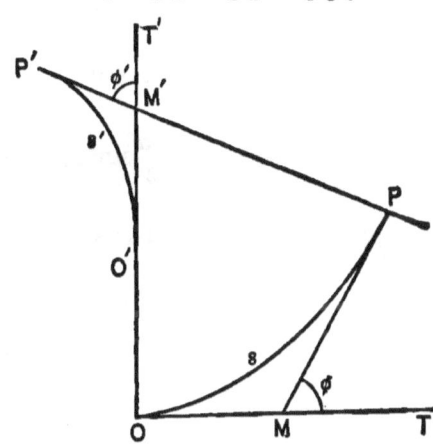

But from (1) $\quad PP' = f'(\phi),$
consequently $\quad OO' = f'(0).$
Hence $\quad s' = f'(\phi) - f'(0) = f'(\phi') - f'(0).$

Omitting the accents on s and ϕ, as no longer necessary, we have, for the intrinsic equation of the evolute,
$$s = f'(\phi) - f'(0).$$

93. For example, from the intrinsic equation of the cycloid
$$s = 4a\sin\phi = f(\phi),$$
we have $\quad f'(\phi) = 4a\cos\phi,$
and $\quad f'(0) = 4a.$

Hence the equation of the evolute is
$$s = 4a(\cos\phi - 1),$$
s being negative, as the radius of curvature is decreasing.

EXAMPLES.

Find the intrinsic equations of the following curves, and of their evolutes.

1. $y = \dfrac{a}{2}(e^{\frac{x}{a}} + e^{-\frac{x}{a}})$. *Ans.* $s = a\tan\phi$, and $s = a\tan^2\phi$.

2. $x^{\frac{2}{3}} + y^{\frac{2}{3}} = a^{\frac{2}{3}}$. *Ans.* $s = \dfrac{3a}{2}\sin^2\phi$, and $s = \dfrac{3a}{2}\sin 2\phi$.

3. $r = a(1 - \cos\theta)$. *Ans.* $s = 4a\operatorname{vers}\dfrac{\phi}{3}$, and $s = \dfrac{4a}{3}\sin\dfrac{\phi}{3}$.

APPENDIX.

ANGLES MADE WITH THE CO-ORDINATE PLANES BY THE TANGENT PLANE OF A SURFACE.

94. The expression for sec γ on page 268 may be derived as follows:

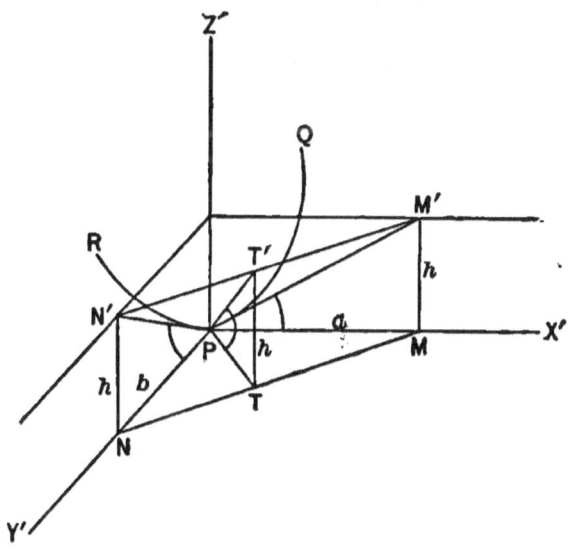

Let P, a point of the given surface, be the point of contact. Through P draw PX', PY', PZ', parallel to the co-ordinate axes.

Let the curve PQ be the section of the surface by the plane $X'Z'$, and PM' tangent to it; also PR the section by the plane $Y'Z'$, and PN' tangent to it.

Since y is constant for points in the plane $X'Z'$, it is evident that the tangent of the angle $M'PX'$ is the *partial* differential coefficient of z with respect to x; that is,

$$\tan M'PX' = \frac{\partial z}{\partial x}.$$

Similarly, $\tan N'PY' = \dfrac{\partial z}{\partial y}.$

As the tangent plane at P contains the two tangent lines PM' and PN', the plane $M'PN'$ is the tangent plane.

Pass a plane parallel to $X'Y'$ at the distance h above it, intersecting the tangent lines in the points M', N', whose projections are M, N.

Draw MN, and PT perpendicular to it, and erect the plane PTT' perpendicular to $X'Y'$.

Then $T'PT = \gamma,$

the angle made by the tangent plane $M'PN'$ with $X'Y'$.

Let $PM = a, \quad PN = b.$

By similar triangles

$$PT : a = b : MN = b : \sqrt{a^2 + b^2},$$

$$PT = \frac{ab}{\sqrt{a^2+b^2}}.$$

$$\tan T'PT = \frac{h}{PT} = \frac{h\sqrt{a^2+b^2}}{ab}.$$

$$\tan^2 T'PT = \frac{h^2}{a^2} + \frac{h^2}{b^2} = \tan^2 M'PM + \tan^2 N'PN,$$

that is, $\tan^2 \gamma = \left(\dfrac{\partial z}{\partial x}\right)^2 + \left(\dfrac{\partial z}{\partial y}\right)^2,$

$$\sec^2 \gamma = 1 + \left(\frac{\partial z}{\partial x}\right)^2 + \left(\frac{\partial z}{\partial y}\right)^2.$$

TANGENT PLANE.

95. *Another Method.*

Let α, β, γ, be the angles made by the normal to the surface at P with PX', PY', PZ'.

Let angles $\quad M'PM = A, \quad N'PN = B.$

The direction cosines of PM' are $\cos A, \quad 0, \quad \sin A;$

of PN', $\qquad\qquad\qquad\quad 0, \quad \cos B, \quad \sin B.$

Since the normal is perpendicular to both PM' and PN', we must have
$$\cos\alpha \cos A + \cos\gamma \sin A = 0,$$
and
$$\cos\beta \cos B + \cos\gamma \sin B = 0,$$
from which
$$\cos\alpha = -\tan A \cos\gamma,$$
$$\cos\beta = -\tan B \cos\gamma.$$

Substituting these expressions in
$$\cos^2\alpha + \cos^2\beta + \cos^2\gamma = 1,$$
we have $\quad \cos^2\gamma (\tan^2 A + \tan^2 B + 1) = 1,$

$$\sec^2\gamma = 1 + \tan^2 A + \tan^2 B = 1 + \left(\frac{\partial z}{\partial x}\right)^2 + \left(\frac{\partial z}{\partial y}\right)^2.$$

$$\sec^2\alpha = \frac{\sec^2\gamma}{\tan^2 A} = \frac{\sec^2\gamma}{\left(\frac{\partial z}{\partial x}\right)^2},$$

$$\sec^2\beta = \frac{\sec^2\gamma}{\tan^2 B} = \frac{\sec^2\gamma}{\left(\frac{\partial z}{\partial y}\right)^2}.$$

ADVERTISEMENTS.

ANALYTIC GEOMETRY

PLANE AND SOLID.

BY E. W. NICHOLS,

Professor of Mathematics in the Virginia Military Institute.

The aim of the author has been to prepare a work for beginners, and at the same time to make it sufficiently comprehensive for the requirements of the usual undergraduate course. For the methods of development of the various principles he has drawn largely upon his experience in the classroom. In the preparation of the work, all authors, home and foreign, whose works were available, have been freely consulted.

In the first few chapters elementary examples follow the discussion of each principle. In the subsequent chapters, sets of examples appear at intervals throughout each chapter, and are so arranged as to partake both of the nature of a review and an extension of the preceding principles. At the end of each chapter general examples, involving a more extended application of the principles deduced, are placed for the benefit of those who may desire a higher course in the subject.

Nichols's Analytic Geometry is in use as the regular text in the greater number of the larger colleges and universities, and has proved itself adapted to the needs of institutions with the most varied requirements.

Cloth. Pages xii + 275. Introduction price, $1.25.

D. C. HEATH & CO., Publishers, Boston, New York, Chicago

NUMBER AND ITS ALGEBRA
BY ARTHUR LEFEVRE, C.E.
Instructor in Pure Mathematics in the University of Texas.

In the form of a syllabus of lectures on the theory of number and its algebra, introductory to a collegiate course in algebra, this monograph presents a thorough exposition of number as conceived and used in mathematics, in form comprehensible by those not already thoroughly versed in the science. It is no mere psychological discussion of mental processes antecedent to the primary concept of number, but the self-consistent development of the concept through phases undreamed of by the man whose sole notion of number is his abstraction from a flock of sheep or pile of coins, — the whole being a body of knowledge essential to right teaching at any stage of systematic mathematical instruction.

Among the universities that have adopted this work are Harvard and the universities of Pennsylvania and Virginia.

Cloth. Pages, iv + 230. Price, $1.25.

THE NUMBER SYSTEM OF ALGEBRA
Treated Theoretically and Historically
BY HENRY B. FINE, PH.D.
Professor of Mathematics in Princeton University

The theoretical part of this book is an elementary exposition of the nature of the number concept, of the positive integer, and of the four artificial forms of number, which, with the positive integer, constitute the "number system" of algebra, viz., the negative, the fraction, the irrational, and the imaginary.

The historical part presents a *résumé* of the history of the most important parts of elementary arithmetic and algebra.

Cloth. Pages, x + 131. Price, $1.00.

D. C. HEATH & CO., Publishers, Boston, New York, Chicago

THEORY OF EQUATIONS

BY SAMUEL MARX BARTON, PH.D.,

Professor of Mathematics in the University of the South.

In this treatise the author aims to give the elements of Determinants and the Theory of Equations in a form suitable, both in amount and quality of matter, for use in undergraduate courses. The work is readily intelligible to the average student who has become proficient in algebra and the elements of trigonometry. The use of the calculus has been purposely avoided. While the presentation of the subject has necessarily been condensed to suit the requirements of college courses, great pains has been taken not to sacrifice clearness to brevity. It is a short treatise, but not a syllabus.

Part I treats of Determinants. The chapters give the fundamental theorems, with examples for illustration; applications and special forms of determinants, followed by a collection of carefully selected examples.

Part II treats of the Theory of Equations proper, with chapters upon complex numbers, properties of polynomials, general properties of equations, relations between roots and coefficients, symmetric functions, transformation of equations, limits of the roots of an equation, separation of roots, elimination, solution of numerical equations. Almost every theorem is elucidated by the complete solution of one or more representative examples.

Cloth. Pages, x + 198. Introduction price, $1.50.

D. C. HEATH & CO., Publishers, Boston, New York, Chicago

COLLEGE ALGEBRA

BY EDWARD A. BOWSER, LL.D.

Professor of Mathematics and Engineering in Rutgers College.

This work is designed for academies, colleges and scientific schools. It begins with the elements, and the full treatment of the earlier parts renders it unnecessary that students who use it shall have previously studied a more elementary algebra.

Among its points of superiority are the following:—

1. Completeness of treatment combined with simplicity.
2. Avoidance of the abstruse and the elaborate in treating the more difficult parts of the subject.
3. Definiteness of statement—the steps and processes are generally formulated in plain rules.
4. Careful consideration and clear presentation of material for the student.
5. Systematic arrangement of material under each subject.
6. Full notes of explanation, direction, and information, useful to student and teacher.
7. Numerous examples are distributed throughout the text in immediate connection with the principles they illustrate.

Half leather. Pages, xviii + 540. Introduction price, $1.50.

Bowser's Academic Algebra, $1.12.
Bowser's Plane and Solid Geometry, $1.25.
Bowser's Elements of Place and Spherical Trigonometry, 90 cents.
    With tables, $1.40.
Bowser's Five Place Logarithmic and Trigonometric Tables, 50 cents.
Bowser's Treatise on Trigonometry, $1.50.

D. C. HEATH & CO., Publishers, Boston, New York, Chicago

COLLEGE ALGEBRA

By WEBSTER WELLS, S.B.,

Professor of Mathematics in the Massachusetts Institute of Technology.

The first eighteen chapters have been arranged with reference to the needs of those who wish to make a review of that portion of Algebra preceding Quadratics. While complete as regards the theoretical parts of the subject, only enough examples are given to furnish a rapid review in the classroom.

Attention is invited to the following particulars on account of which the book may justly claim superior merit: —

The proofs of the five fundamental laws of Algebra — the Commutative and Associative Laws for Addition and Multiplication, and the Distributive Law for Multiplication — for positive or negative integers, and positive or negative fractions; the proofs of the fundamental laws of Algebra for irrational numbers; the proof of the Binomial Theorem for positive integral exponents and for fractional and negative exponents; the proof of Descartes's Rule of Signs for Positive Roots, for incomplete as well as complete equations; the Graphical Representation of Functions; the solution of Cubic and Biquadratic Equations.

In Appendix I will be found graphical demonstrations of the fundamental laws of Algebra for pure imaginary and complex numbers; and in Appendix II, Cauchy's proof that every equation has a root.

Half leather. Pages, vi + 578. Introduction price, $1.50.
Part II, beginning with Quadratics. 341 pages. Introduction price, $1.32.

D. C. HEATH & CO., Publishers, Boston, New York, Chicago

TREATISE ON TRIGONOMETRY
AND ITS APPLICATIONS TO
ASTRONOMY AND GEODESY

By EDWARD A. BOWSER, LL.D.
Professor of Mathematics and Engineering in Rutgers College

The aim of the author has been to present in as concise a form as is consistent with clearness, the fullest course in Trigonometry which is given in the best technical schools and in advanced courses in colleges.

The examples are very numerous and are carefully selected. Among these are some of the most elegant theorems in Plane and Spherical Trigonometry. The numerical solution of triangles has received much attention, each case being treated in detail.

The chapters on De Moivre's Theorem, and Astronomy, Geodesy, and Polyhedrons will serve to introduce the students to some of the higher applications of Trigonometry, rarely found in American text-books.

American Mathematical Monthly: Excepting one, this is the most complete Treatise on Trigonometry published in America, and in point of excellence is superior to that work. In the method of treatment, arrangement, typographical execution, and numerous and well-selected exercises, it has no superior. The definitions of the functions are given "once for all" and need not be restated and modified when obtuse and reflex angles are considered.

In the development of the theoretical part of the subject, the work is especially interesting and clear. From the beginning the student is carried along with enthusiasm and with the assurance that he is mastering the subject. The unusually large and well-chosen collection of problems are suited to every requirement, and by solving these the student learns to do by doing.

The treatment of Trigonometric Elimination, De Moivre's Theorem, Summation of Series, etc., is more complete than is usually given in text-books.

These observations have been gathered by using the book in the class-room.

Half leather. Pages, xiv + 368. Introduction price, $1.50.

Bowser's Five Place Logarithmic and Trigonometric Tables, 50 cents.
Bowser's Elements of Plane and Spherical Trigonometry, 90 cents. With tables, $1.40.
Bowser's Plane and Solid Geometry, $1.25.
Bowser's Academic Algebra, $1.12.
Bowser's College Algebra, $1.50.

D. C. HEATH & CO., Publishers, Boston, New York, Chicago

SURVEYING AND NAVIGATION
AN ELEMENTARY TREATISE
BY ARTHUR G. ROBBINS, S.B.,
Assistant Professor of Civil Engineering, Massachusetts Institute of Technology

This brief work on Surveying and Navigation is intended for those students who desire to supplement the study of Trigonometry with a brief course in its applications to those subjects. Although the subjects are treated in an elementary way, the work is correct and accurate as far as it goes.

Paper. Pages, vi + 61. Price, 50 cents.
Bound with Wells's New Plane and Spherical Trigonometry, with tables, $1.50.

PLANE AND SPHERICAL TRIGONOMETRY
BY E. MILLER, A.M.
Professor of Mathematics and Astronomy in the University of Kansas.

The work is arranged in a clear and logical order. It brings the principles of the subject face to face with operations, and thus not only satisfies the student of the mutual dependence of the two, but tends to carry him back to a clear apprehension of what he had probably failed to appreciate in the subordinate sciences. It has been remarked by many who have used the work as a text, that it contains the best exposition and development of Spherical Trigonometry ever published.

Cloth. Pages, vi + 114. With tables, $1.40.

D. C. HEATH & CO., Publishers, Boston, New York, Chicago

DESCRIPTIVE GEOMETRY
By CLARENCE A. WALDO, Ph.D.
Professor of Mathematics in Purdue University

The special features of this work are: The method of developing the subject by problems systematically arranged, and supplemented by suggestions when needed; the large number of problems given; the method of stating the problems, which, in connection with the notation adopted, makes every lettered drawing entirely self-explanatory; the introduction of several subjects of considerable descriptive value, such as the axis of affinity, axonometry, Pascal's and Brianchon's hexagons; the early discussion of the cone and cylinder of revolution and the sphere, in order that from the beginning these surfaces may be used as auxiliary; the omission of all plates except a few of a generic character.

Cloth. Pages, xii + 77. Price, 80 cents.

GEOMETRICAL TREATMENT OF CURVES
Which are isogonal conjugate to a straight line with respect to a triangle
By ISAAC J. SCHWATT, Ph.D.
Assistant Professor of Mathematics in the University of Pennsylvania

The discussion includes the hyperbola and several aspects of the ellipse. Three large folding plates illustrate the application of principles.

Paper. Octavo. Pages, iv + 45. Price, $1.00.

CONIC SECTIONS
By RUFUS B. HOWLAND, B.C.E.
Professor of Mathematics in Wyoming Seminary, Kingston, Pa.

This manual presents the elements of Conic Sections in a form suited to the capacity of advanced classes in Geometry.

Cloth. Pages, iv + 60. Price, 75 cents.

D. C. HEATH & CO., Publishers, Boston, New York, Chicago

Science.

Ballard's World of Matter. A guide to mineralogy and chemistry. $1.00.

Benton's Guide to General Chemistry. A manual for the laboratory. 35 cents.

Boyer's Laboratory Manual in Biology. An elementary guide to the laboratory study of animals and plants. 80 cents.

Boynton, Morse and Watson's Laboratory Manual in Chemistry. 50 cents.

Chute's Physical Laboratory Manual. A well-balanced course in laboratory physics, requiring inexpensive apparatus. Illustrated. 80 cents.

Chute's Practical Physics. For high scnools and colleges. $1.12.

Clark's Methods in Microscopy. Detailed descriptions of successful methods. $1.60.

Coit's Chemical Arithmetic. With a short system of analysis. 50 cents.

Colton's Physiology: Experimental and Descriptive. For high schools and colleges. Illustrated. $1.12.

Colton's Physiology: Briefer Course. For earlier years in high schools. Illustrated. 90 cents.

Colton's Practical Zoology. Gives careful study to typical animals. 60 cents.

Grabfield and Burns's Chemical Problems. For review and drill. Paper. 25 cents.

Hyatt's Insecta. A practical manual for students and teachers. Illustrated. $1.25.

Newell's Experimental Chemistry. A modern text-book in chemistry for high schools and colleges. $1.10.

Orndorff's Laboratory Manual. Contains directions for a course of experiments in Organic Chemistry, arranged to accompany Remsen's Chemistry. Boards. 35 cents.

Pepoon, Mitchell and Maxwell's Plant Life. A laboratory guide. 50 cents.

Remsen's Organic Chemistry. An introduction to the study of the compounds of carbon. For students of the pure science, or its application to arts. $1.20.

Roberts's Stereo-Chemistry. Its development and present aspects. 1.00.

Sanford's Experimental Psychology. Part I. Sensation and Perception. $1.50.

Shaler's First Book in Geology. Cloth, 60 cents. Boards, 45 cents.

Shepard's Inorganic Chemistry. Descriptive and qualitative; experimental and inductive; leads the student to observe and think. For high schools and colleges. $1.12.

Shepard's Briefer Course in Chemistry, with chapter on Organic Chemistry. For schools giving a half year or less to the subject, and schools limited in laboratory facilities. 80 cents.

Shepard's Laboratory Note-Book. Blanks for experiments; tables for the reactions of metallic salts. Can be used with any chemistry. Boards. 35 cents.

Spalding's Botany. Practical exercises in the study of plants. 80 cents.

Stevens's Chemistry Note-Book. Laboratory sheets and covers. 50 cents.

Venable's Short History of Chemistry. For students and the general reader. $1.00.

Walter, Whitney and Lucas's Animal Life. A laboratory guide. 50 cents.

Whiting's Physical Measurement. I. Density, Heat, Light, and Sound. II. Dynamics, Magnetism, Electricity. III. Principles and Methods of Physical Measurement, Physical Laws and Principles, and Tables. Parts I-IV, in one volume, $3.75.

Whiting's Mathematical and Physical Tables. Paper. 50 cents.

Williams's Modern Petrography. Paper. 25 cents.

For elementary works see our list of books in Elementary Science.

D.C. HEATH & CO., Publishers, Boston, New York, Chicago

Mathematics

Barton's Theory of Equations. A treatise for college classes. $1.50.

Bowser's Academic Algebra. For secondary schools. $1.12.

Bowser's College Algebra. A full treatment of elementary and advanced topics. $1.50.

Bowser's Plane and Solid Geometry. $1.25. PLANE, bound separately. 75 cts.

Bowser's Elements of Plane and Spherical Trigonometry. 90 cts.; with tables, $1.40.

Bowser's Treatise on Plane and Spherical Trigonometry. An advanced work for colleges and technical schools. $1.50.

Bowser's Five-Place Logarithmic Tables. 50 cts.

Fine's Number System in Algebra. Theoretical and historical. $1.00.

Gilbert's Algebra Lessons. Three numbers: No. 1, to Fractional Equations; No. 2. through Quadratic Equations; No. 3, Higher Algebra. Each number, per dozen, $1.44.

Hopkins's Plane Geometry. Follows the inductive method. 75 cts.

Howland's Elements of the Conic Sections. 75 cts.

Lefevre's Number and its Algebra. Introductory to college courses in algebra. $1.25.

Lyman's Geometry Exercises. Supplementary work for drill. Per dozen, $1.60.

McCurdy's Exercise Book in Algebra. A thorough drill book. 60 cts.

Miller's Plane and Spherical Trigonometry. For colleges and technical schools. $1.15. With six-place tables, $1.40.

Nichol's Analytic Geometry. A treatise for college courses. $1.25.

Nichols's Calculus. Differential and Integral. $2.00.

Osborne's Differential and Integral Calculus. $2.00.

Peterson and Baldwin's Problems in Algebra. For texts and reviews. 30 cts.

Robbins's Surveying and Navigation. A brief and practical treatise. 50 cts.

Schwatt's Geometrical Treatment of Curves. $1.00.

Waldo's Descriptive Geometry. A large number of problems systematically arranged and with suggestions. 80 cts.

Wells's Academic Arithmetic. With or without answers. $1.00.

Wells's Essentials of Algebra. For secondary schools. $1.10.

Wells's Academic Algebra. With or without answers. $1.08.

Wells's New Higher Algebra. For schools and colleges. $1.32.

Wells's Higher Algebra. $1.32.

Wells's University Algebra. Octavo. $1.50.

Wells's College Algebra. $1.50. Part II, beginning with quadratics. $1.32.

Wells's Essentials of Geometry. (1899.) $1.25. PLANE, 75 cts. SOLID, 75 cts.

Wells's Elements of Geometry. *Revised.* (1894.) $1.25. PLANE, 75 cts.; SOLID, 75 cts.

Wells's New Plane and Spherical Trigonometry. For colleges and technical schools. $1.00. With six place tables, $1.25. With Robbins's Surveying and Navigation, $1.50.

Wells's Complete Trigonometry. Plane and Spherical. 90 cts. With tables, $1.08. PLANE, bound separately, 75 cts.

Wells's New Six-Place Logarithmic Tables. 60 cts.

Wells's Four-Place Tables. 25 cts.

For Arithmetics see our list of books in Elementary Mathematics.

D. C. HEATH & CO., Publishers, Boston, New York, Chicago

www.ingramcontent.com/pod-product-compliance
Lightning Source LLC
Chambersburg PA
CBHW022019240426
43667CB00042B/944